Machine Tools - Design, Research, Application

Edited by Ľubomír Šooš and Jiri Marek

Published in London, United Kingdom

IntechOpen

Supporting open minds since 2005

Machine Tools - Design, Research, Application
http://dx.doi.org/10.5772/intechopen.83266
Edited by Ľubomír Šooš and Jiri Marek

Contributors
Oleg Krol, Karlo Obrovac, Toma Udiljak, Jadranka Vukovic Obrovac, Miho Klaic, Jozef Svetlík, Tomas Stejskal, Peter Demec, Ľubomír Šooš, Michal Holub, Jiri Marek, Tomas Marek, Petr Blecha, Jiří Tůma, Miroslav Mahdal, Jiří Šimek, Jaromír Škuta, Renata Wagnerová, Stanislav Žiaran

First published in London, United Kingdom, 2020 by IntechOpen
IntechOpen is the global imprint of INTECHOPEN LIMITED, registered in England and Wales, registration number: 11086078, 5 Princes Gate Court, London, SW7 2QJ, United Kingdom
Printed in Croatia

British Library Cataloguing-in-Publication Data
A catalogue record for this book is available from the British Library

Additional hard and PDF copies can be obtained from orders@intechopen.com

Machine Tools - Design, Research, Application
Edited by Ľubomír Šooš and Jiri Marek
p. cm.
Print ISBN 978-1-83962-350-9
Online ISBN 978-1-83962-351-6
eBook (PDF) ISBN 978-1-83962-352-3

For EU product safety concerns:
IN TECH d.o.o., Prolaz Marije Krucifikse Kozulić 3, 51000 Rijeka, Croatia, info@intechopen.com or visit our website at intechopen.com.

Meet the editors

Ľubomír Šooš is currently the Dean of the Faculty of Mechanical Engineering at the Slovak University of Technology in Bratislava. His professional background is in the theory and design, especially of machine tools and equipment for machinery production. Until now he has published more than 295 scientific articles in journals and is the author of 52 national and international patents. As a leader he participated on 49 international and national research projects. He is a member of several editorial boards of the international scientific journal "Manufacturing Technology", "Praise Worthy Prize", "MM Science", "Waste Management Forum" and others. Professor Šooš has achieved very good collaboration with the industry. He is also Vice-President of the Automotive Industry Association of the Slovak Republic. For his excellent cooperation he received the "Award of Rector TU" Novi Sad (Srb), Award of FME CVUT(CZ), Medal of Georgia Agricoly in Ostrava (CZ), "Professor of the year 2015" (SK) and his Doctor honoris causa on TU Ostrava (CZ).

Professor Marek worked as a Technical Director at TOSHULIN and TOS Kuřim - OS company from 2000 to 2016. Now he is a lecturer at the Technical University in Brno, Faculty of Mechanical Engineering, Institute of Production Machines, Systems and Robotics of Engineering. In his practical and pedagogical activities, he aims especially at design of machine tools and machining centres for rotary and non-rotary workpieces, at theory of the designing process and of the product life cycle, and last but not least, at system methodology. He is a member of many professional institutions and works as a member of the managing committee in the scientific section of the magazine MM Science Journal.

Contents

Preface

You have just opened the book "Machine tools - design, research and applications". Machine tools are systems designed to create workpieces of a particular shape, dimension and machining quality. For this purpose machine tools have to create cutting motions that consist of the mutual coupling of vectors of rotary motion and translational motion. Today, we encounter machine tools everywhere and we cannot imagine life without them.

An important role in the development of today's type of machine tools was played by the rolling bearing patent obtained by the Englishman Philip Vaughan in 1794 and, three years later, by the rediscovery of the lathe by his compatriot Henry Maudslay.

The development and application of new, highly productive cutting tools made of cutting ceramics, natural or synthetic diamonds and other synthetic super hard materials allowed a significant increase in the values of cutting speeds. The pioneer of high-speed machining is considered to be the German researcher Carl Salamon, who in 1920 milled steel at the cutting speed of 440 $m.min^{-1}$, and aluminium at up to 16 500 $m.min^{-1}$. This significantly reduced overall machining times, increased machine productivity and production efficiency.

The productivity of the machine tool is characterized by the number of manufactured parts along with the size of the machined area and the volume of material used. Productivity is limited by the mean thickness of the chips and it depends on cutting and feed speeds. Moreover, the cutting speed depends mostly on the frequency of rotation of the headstock of the machine tool. This is the reason for the ongoing increase in mean revolution frequencies, the ever more frequent application of headstocks with integrated drive, the so-called "Electro-spindle" and the shift from classic to high-speed machining.

In proportion to the growth of living standards, the speed, quantity and variety of requirements for the quality and accuracy of manufactured products have increased. The means to achieve the required accuracy of the machine tool is by optimizing the structure in terms of stiffness and dynamic stability.

Therefore, in addition to increasing the productivity of machine tools, it is of the utmost importance to pay special attention to optimizing the design of the machine in terms of rigidity, thermal expansion, variability and chip removal at large volumes of machined material. Therefore, when designing and optimizing the construction of machines, we increasingly resort to virtual prototyping and computer analysis of the functional properties of the proposed machine.

The basic condition for achieving the required machine tool accuracy, as well as that of the machined parts, is the rigidity of the "machine - tool - preparation - workpiece" system. Apart from rigidity, this system is influenced by the geometric accuracy of the machine, the technological approach, the strategy for measuring the work piece and the servicing of the machine. With classical machine tool designs, this refers to the serial structure of arranging the motion axes, where the total rigidity of the system is limited by the machine's weakest construction node. In machine tools with moving rotation, this generally is the motion axis carrying the workpiece, or the tool carrier - against the headstock.

It is clear that such productivity and accuracy of machine tools depends mostly on the quality of the headstock and other structural elements, such as guide parts, drives, frame and other nodes. Both of these criteria act in a contradictory manner. With the requirement for increased headstock frequency and feed speed, production does increase. However, at the same time the positional rigidity of the headstock and of the motion mechanisms – in other words, machine work precision – decreases. From the viewpoint of productivity and accuracy of work we can regard the headstock as the heart of the whole machine tool, influencing the quality of work to a decisive degree.

Among the other important quality criteria is the mechanical design or modularity of the machine tool, or of the production system. The modular state of the construction is an important precondition for the competitiveness of the product on the market. The basis of this concept is the system of modules from which various configurations can be created, in line with the specific demands of the client. On the part of the producer, the manufacture of the machines will be made more effective, while, on the other hand, the client will be paying for a functional machine that he really needs and uses.

An inseparable component in the selection of a suitable machine is risk management and a current analysis of the functionality and reliability of the machine's work. Based on this analysis, we are able to continue with a computer optimization of the mutual spatial arrangements of the configuration "machine - tool - preparation - workpiece" system to achieve the maximum work safety of the existing machine.

In recent times, it has become a matter of course to include, within the added value for the mechanical and electronic construction of the CNC machining tool, the capacity to compensate for inaccuracy over the whole working area – so-called volumetric compensation. Indeed, it has been shown that contemporary mechanical construction has reached its potential limits in the current period, unless some new principle is discovered. For this reason, in addition to heat stabilization of the machine, one of the alternatives for increasing precision of work is to increase volumetric precision through spatial compensations. This will require familiarity with the perfect geometric precision of the tool, which is influenced by the manufacture of the individual pieces of the motion axis and their assembly, involving all the axes at once.

This book is divided into three basic sections and seven chapters presenting both theoretical and experimental work. Chapters 1-3 cover trends in the development of machine tool construction. Chapters 4-6 deal with the research and development of system rigidity, and the measuring of the accuracy of the machine tools.

Chapter 7 presents an example of the application of machine tools in the production of orthopaedic aids. The work presented in the book is of considerable relevance and use to researchers working in the area of design, research and development.

Dr.h.c. prof. Ing Ľubomír Šooš, PhD.
Dean,
Faculty of Mechanical Engineering,
Slovak University of Technology Bratislava,
Slovakia

Jiri Marek
Professor,
Brno University of Technology,
Czechia

Section 1

Design and Trend

Chapter 1

Modularity of Production Systems

Jozef Svetlík

Abstract

From the theoretical point of view, the chapter focuses on the unification of views on the living (constantly changing) structure of the construction of flexible production systems, including its cooperating devices. It contains currently defined and designated technical terms in the field of flexible production systems. From the theoretical point of view, the existing structures of the "multiprofessional manufacturing robotic center" are enhanced with new elements, which also contributes to innovation and expansion of their applications. These structural structures served as the basis for building sophisticated modular structures. Modularity is an integrating element directed at highly customizable manufacturing engineering structures. It fully complies with the requirements of manufacturing practice and demanding market, in the framework of fully implemented Industry 4.0 (I4.0) under way.

Keywords: modularity, module, production systems, structure, platform

1. Introduction

Modular manufacturing systems, as an integrated part of flexible manufacturing systems, deserve an unmistakable merit in today's rapidly changing manufacturing environment, characterized by developed competition in the global context and progressive changes in process technologies and in their structure according to market requirements. Such systems necessitate a rapid and factual integration of new technologies and new functions into both system and process relationships.

The Industry 4.0 (I4.0) trends and conditions and requirements require cyber and flexible production-oriented approach, enabling to build the following:

- A production capacity of production systems that is operatively adaptive to market requirements, i.e., obtaining new, rapidly viable products

- Fast integration of modern process technologies and new functions into existing production systems and their easy adaptation to dynamically changing batches of individual products

- Integrated production units with new service capabilities based on robust Industry Internet of Things (IIoT) data streams from individual work units and their accessibility for being processed from anywhere subject to Internet connection

2. Flexible manufacturing systems

Flexible manufacturing systems (FMS) enable flexible production of a product group in a single production system. Using modular principles, flexible manufacturing, which is the fundamental concept of cyber production systems, has recently become one of the major systems of production management. These arrangements are (and there are several of them) theoretically and methodically based on the search for a mathematically modeled component production center relationship that would guarantee different types of parts produced with a small number of pieces in the batch. The modular structure of the production systems enables links between machines, saving production time and space. The operation of the machines is synchronized via data stream, and the material flow is optimized (moving parts between machines is at an optimal distance). FMS utilizes many advantages of other types of production structures (**Table 1**) [1].

The dynamic development of computers, information science, data processing, control and managing systems, optical systems, drives, and materials, that is taking place in short cycles, significantly affects the growth rate (obsolescence) of the technical level of the systems in question. An efficient manufacturing system can become inefficient in a short time. In addition, the current customer-oriented market, as well as the environmental, energy and material issues, results in accelerated launch of new products. The adaptability of established manufacturing systems to new products may not have sufficient technical availability, and the introduction

Type of production	Structure definitions and objectives
Production line	The line is designed for the production of a (one) specific product, using the technology of gradual production with given tools and a fixed level of automation. The economic goal of production lines is to produce one particular type of product in large quantities and the required quality cost-effectively
Flexible manufacturing system (**FMS**)	The structure of a production system with fixed hardware and programmable software for affecting changes in the assortment produced according to current orders and changes in production plans with tools for several product types. The economic objective of **FMS** is to ensure an efficient production of several types of products, which may change over time with the respective changes taking up shorter time on the same production system, while maintaining the requirements for the production's prescribed scope and quality.
Reconfigurable manufacturing system (**RMS**)	A structure of the manufacturing system that can be created through multiple groupings of basic process configurations of changeable system modules (hardware and software). Reconfiguration allows for the addition, removal, or modification of specific process features, controls, control software, or machine structure to adapt the system's production capacity to changes in market demand or to the necessary and related technological changes. This system structure guarantees the flexibility of the system for a specific product group, while the system is technically ready for change so that it can be further improved, upgraded, and reconfigured and not merely replaced [2, 3]. The goal of **RMS** is to provide the functionality and capacity that is needed at any given time. In terms of the system composition, **RMS** configuration can be reserved or flexible or changeable as needed between these two properties. The **RMS** goals exceeds the **FMS** goals in terms of economy, allowing the following: • Shortening the time of introduction of innovative systems and reconfiguration of the existing ones • Immediate production adjustment and rapid integration of new technologies and new functions into existing production systems

Table 1.
Overview of basic production system structures.

of new technically available systems may take too long a time from the production availability point of view (machine tools approximately 2 years).

For these reasons, it is necessary to pay constant attention to flexible, modular, and reconfigurable production systems and consequently to improve them systematically and technically and adapt them to the needs of current production processes or to the needs of current engineering production [1, 2].

Generally, the best-selling article (or article with the highest investment value) of production technology are the CNC machine tools. Prof. Marek writes in [3] about the factors influencing the development of machine tools.

Forecasts focused on the position of modular technologies in the twenty-first century confirm their important place in both fully automated production plants (both engineering and nonengineering areas) and in non-production areas (service and maintenance activities).

Thus, the modularity and reconfigurability in terms of where the development is heading have the potential for further development in the future. Design of reliable (universal) modules or of the building nodes with a wider applicability is, and will always be, topical. This desired property can be achieved through experience, selection of suitable elements, and reliable design. In terms of reliability and reconfigurability, two areas need to be focused on:

- Design of machine tools (machine reconfiguration to another type of workpiece)

- Production (technology reconfiguration to another type of workpiece)

The concept of flexibility is also related to reconfigurability and structurelability. Flexibility can also be seen from the point of view of design and manufacturing development (**Figure 1**).

The possibilities and tools for increasing the performance of production machines in multifunction machinery are associated with the developed ability to fully perform several types of machining, e.g., turning and milling at the same time

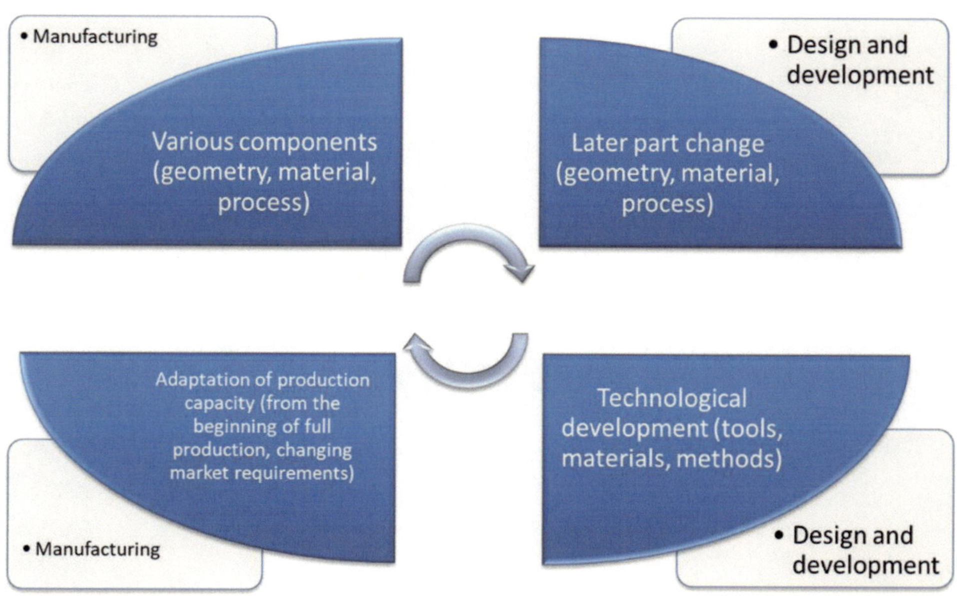

Figure 1.
An unconventional view of the "flexibility" concept.

or milling and grinding, etc. Reducing the number of machine tools for the production of one component, less handling, shortening the lead times, minimizing the recurrence of workpiece clamping, maximizing the concurrence of operations, as well as the development of machine components and machine concepts for maximum machine multifunctionality contribute to:

- Increased accuracy
- Increased production capacity
- Increased economy
- Reduced negative impacts on the environment

Unification of parts and components is implemented in order to minimize diversity of the components used, while maintaining very good static and dynamic properties of the machines and, at the same time:

- Increasing reliability
- Increasing economy

2.1 New approach to production systems classification

In the area of production systems, a number of terms are used with broader interpretation. This situation is related to approaches to and perspectives on this issue. A proposal for their general unification and effective classification is given in **Figure 2**.

Various definitions of production systems from different points of view are cited in various literature sources [1]. This has led to the need to harmonize these formulations so that they provide the most precise definitions, taking into account current knowledge in this area:

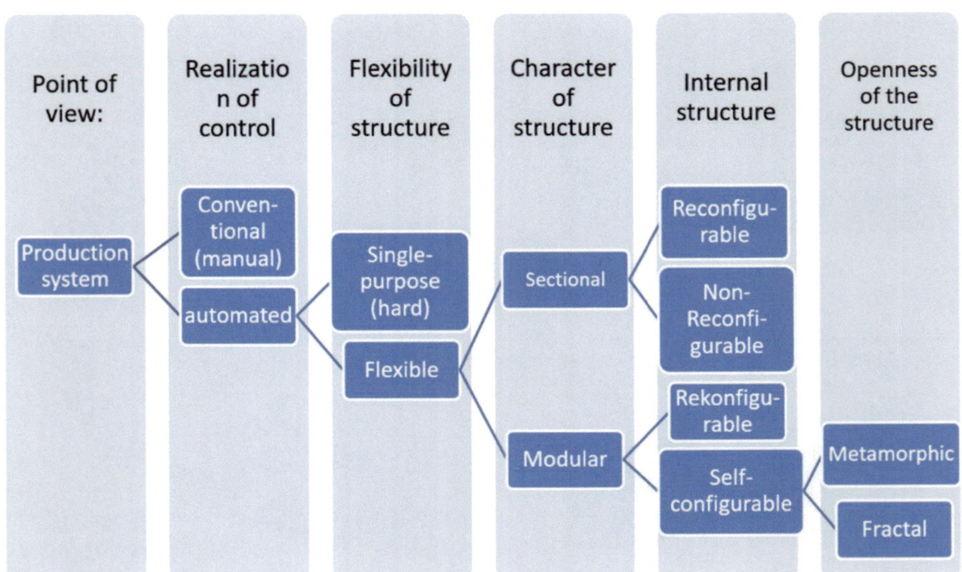

Figure 2.
Open proposal for production systems classification.

Flexible production system—a functional grouping of production facilities linked by material flow and information network, enabling the use of flexible change in production facilities due to the introduction of new products in relatively short time intervals, to produce small quantities efficiently.

Structural production system—flexible set of compatible elements (technological and positioning units, supporting frame, cooling system, etc.) and their mutual links, which can be expanded with new elements to change the system parameters.

Modular system—flexible set of unified modules (module—separately functional unit) in functionally logical (in terms of structure, system, concept, kinematics, etc.) arrangement into higher functional unit (meeting required parameters and working functions).

Reconfigurable system—a modular system with the possibility to change the arrangement of its own modules (in terms of structure, system, concept, kinematics, etc.) in order to create an innovated system with innovated properties.

Self-reconfigurable system—a reconfigurable system capable of independently reconfiguring its own modules (in terms of structure, system, concept, kinematics, etc.) to create an innovated system with innovated features.

Metamorphic system—a closed self-reconfigurable system to create an innovated system with innovated features (def. Inspired by [4]).

Fractal system—an open, self-reconfigurable system consisting of proactively behaving elements—fractals (their structure is recurrent) which pursue a common goal (def. Inspired by [5]).

2.2 Modular production centers

In the category of manufacturing technology, machining centers (MC) are defined as manufacturing machines designated for complex components machining with defined characteristics. According to the number and type of technological operations performed, machining centers are divided into:

- Multipurpose (multi-operation)—machines with a predominant technological operation (e.g., turning), i.e., they mostly enable one type of technology

- Multiprofessional (multiprofessional productions center (MPC))—machines on which various technological operations can be performed (e.g., turning, milling, drilling, etc.)

Production centers are conceptually built on the principles of modular systems or as modular single-purpose machines. In terms of design and structure, they are assembled from technological, handling, and auxiliary units (mechanical, electromechanical, hydraulic, pneumatic) integrated through a supporting element (frame) into one functional and structural unit. The highest integration of production centers is based on the automation of technological and handling operations. These are multiprofessional machine tools designed for complex machining of parts on one machine and, if possible, requiring one clamping. To machine a workpiece requiring one clamping, its rotation must be ensured (e.g., in the X-Y plane) and so must be its tilting. The machining centers are equipped with a tool magazine automatically replaced by a mechanical hand. Some tools feature their own drive, which makes drilling off the workpiece axis or its milling possible, especially on lathes. Machining centers represent the basic AVS production machines. They are mainly used in piece and small batch production. Machining centers are characterized by a high concentration of operations. The machining is often carried out with the component clamped only once. They are mostly equipped with tool magazines exchanged automatically as needed. The most common main feature of machining centers is the largest machined part dimension.

2.3 MPC and MPRC definitions

MPC—a set of working units (technological, handling, conveyors) integrated into one unit (frame), characterized by flexibly reprogrammable common control system, mostly with human operation. The nature and structure of the MPC construction classifies it under the group of modular reconfigurable production systems [1].

MPRC—a fully automated set of autonomous modules, integrated into a single unit (frame), with a common, flexibly reprogrammable control and the use of robotic devices performing the function of handling and technological work units [1, 6]. The nature and structure of the MPRC construction classifies it under the modular reconfigurable production systems.

MPRC characteristic features:

- The MPRC ensures the technological cycle of product manufacturing, i.e., MPRC may be considered a production system.

- The MPRC's design guarantees the automation of technological, handling, and control operations, i.e., MPRC may be deemed an autonomous automated production system.

- The MPRC is built on the principles of modularity, which allows the MPRC to be converted quickly and efficiently into a new product range, i.e., MPRC can be considered a flexible autonomous automated production system.

- By its structure, the MPRC is closest to that of the robotic cell.

Unlike the type-specific structures of automated production systems, the MPRC provides a fully automated multiprofessional technology cycle designed for complete workpiece production and has a simpler (fewer number of elements/modules) structure, less space requirements, and more integration of technological (handling) control functions.

3. Modularity and flexibility of production systems

The technical system is described by terminology which determines the procedures, tools, and methods for its description, understanding, and interpretation.

System—a purposefully defined set of at least two elements and a set of links between them, both sets specifying the properties of the whole. The links can be understood in terms of their physical or logical relationship. From the technical point of view, a system may be mechanical or functional. Each system is made up of individual elements. An important feature of the system is that its elements in relation to each other can work together as a whole. The manifestation and properties of the system represent more than a simple sum of the properties of its elements. The system as a whole may exhibit behavior that is missing in the behavior of its elements.

Subsystem—part of the system, which creates a relatively closed, separate functional unit within the system. As a rule, it consists of two components: elements and links. The links between elements are often called interfaces. The subsystems cooperate with each other in a system function algorithm. The subsystems can be viewed independently.

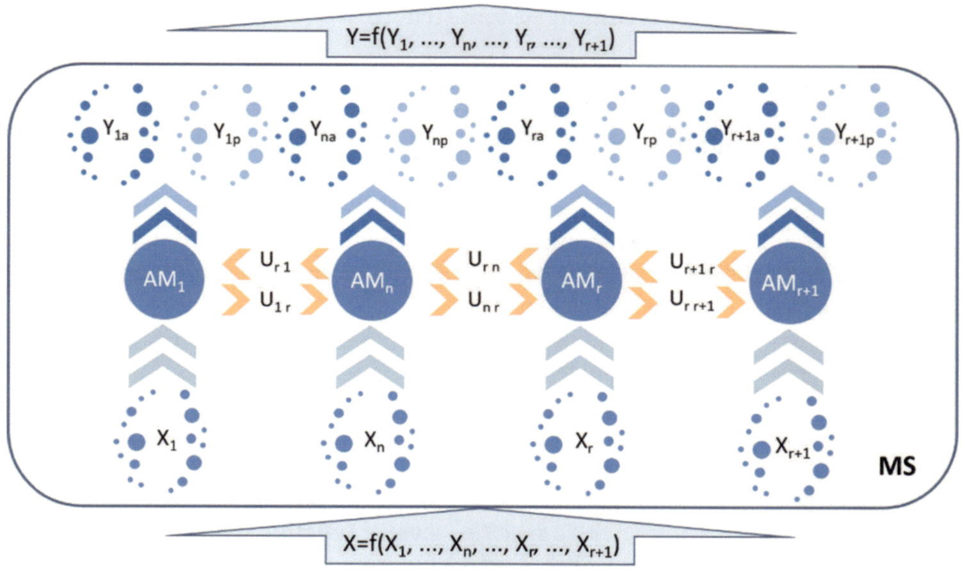

Figure 3.
Structure diagram of the general modular technical system.

Module—a basic building block of modular structure, which is a separate unit structurally, functionally, and in terms of design (a materialized implementation of the basic system function).

Modular system (MS)—complete set of modules (unified units, functional nodes, modular blocks, etc.) in functionally logical (in terms of structure, system, concept, kinematics, etc.) arrangement of a higher functional unit, featuring required parameters and working functions (**Figure 3**) [1].

Modularity—a feature of a technical system that allows its decomposition into a group of autonomous, loosely coupled elements—modules.

Modular system structure—a set of modules and their mutual links and configurations. Change in the interconnection/arrangement of modules results in the emergence of new delimited functional and kinematic system configurations.

Element designation:

MTS—Modular technical system.

AMi—Unified modular unit (functional node, modular block, etc.)

Ui—Mutual links (compatibility of ij module U_i to U_j, or of the ji module U_j to U_i).

X $_i$—Module input parameters (set requirements).

Yi—Module output parameters (properties, operating functions), "a" active, "p" passive.

3.1 Functionality of the modular technical system structure

The mutual linking of AM modules is based on their arrangement in the technical structure of the system ψ. The possibilities of connecting the AM_i and AM_j modules are described by the matrix of the MTS_f system structure (matrix of the type n × n, where n is the number of modules AM = {AM1, AM2, ...,} AMn forming the MTS, while the set of binary links x = {x_{11}, x_{12}, ...,x_{nn}} on the set AM expresses x_{ij} = 1, if there is a possibility of creating a link between AM_i and AM_j, or x_{ij} = 0, if there is no possibility of linking the modules):

$$MTS_f = [x_{ij}] \tag{1}$$

3.2 Assembly of modular technical system structure

By combining the modules AM ={AM_1, AM_2,..., AM_n,}, the MTS can be assembled with none or several degrees of freedom of motion. MTS motion options with respect to a defined coordinate system can be analyzed from the MTS_{pb} motion matrix (n x n matrix type, where n is the number of modules AM = {AM1, AM2,...,AMn,} forming the MTS, where $b_{ij} = 0$ if there is no connection between AM_i and AM_j, or $b_{ij} = 0$ if the modules are combined to form a unit without motion options, or $b_{ij} = 1$ if the modules are combined to form a unit with one degree of freedom of motion, $b_{ij} = 2$ with two degrees, etc.).

$$MTS_{pb} = [b_{ij}] \tag{2}$$

The AM module, a critical element of the MTS ψ structure, is defined as a unified unit, separate structurally, functionally, and in terms of design, composed of elements, elements E (e.g., mechanical module, servo drive, possibly also source, control, and communication module), with a specified level of function integration (main, secondary, auxiliary) and intelligence (control and information, control and decision function), capable of connecting with other modules mechanically, and in terms of control, creating functionally higher units in the technical structure of the system ψ:

$$MTS_{\psi} \approx \sum_{j=1}^{a} AM_j \approx \sum_{j=1}^{a}\sum_{i=1}^{e_j} E_{i,j} \tag{3}$$

AM_{r+1} inputs are X parameters of MTS task transformed to X_{r+1} parameters of X_{r+1} partial task and U_{rr+1} compatibility parameters transformed as interaction of directly linked downstream AM_r module in MTS structure. Outputs from the AM_{r+1} module are output parameters Y_{r+1u} and Y_{r+1p} of the AM_{r+1} module representing the performance of a partial task of the module transformed into output parameters Y of the MTS robot and the U_{r+1r} compatibility parameters, by which the AM_{r+1} module directly affects the subsequently linked module AM_r to the MTS structure (**Figure 4**).

$$X = f(X_1, \ldots, X_n, \ldots, X_r, \ldots, X_{r+1}) \tag{4}$$

$$Y = f(Y_{1u}, \ldots, Y_{nu}, \ldots, Y_{ru}, \ldots, Y_{r+1u}, \ldots, Y_{1p}, \ldots, Y_{np}, \ldots, Y_{r+1p}) \tag{5}$$

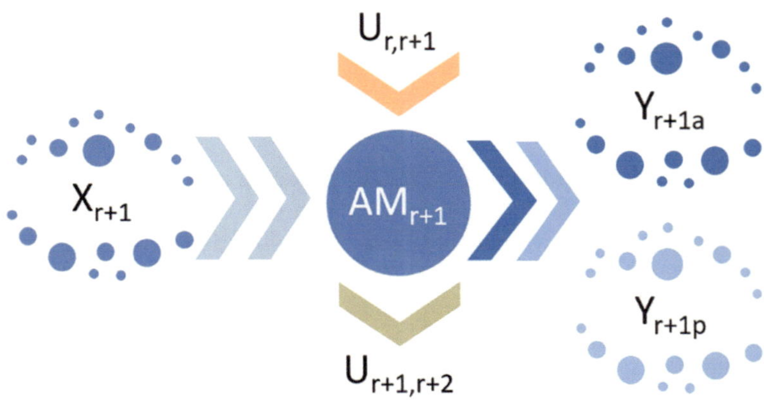

Figure 4.
Structure diagram of the general autonomous module (AM).

AM modules—Depending on their importance for the MTS functions, they can be classified as the main active ones (ensure the main function, number l of total number m + l + a modules, e.g., motion modules), the secondary active/passive ones (ensure secondary support function, remaining number a out of total number m + l + a modules, e.g., a coupler module), and the auxiliary passive ones (ensure the auxiliary function, remaining number a out of the total number of m + l + a modules, e.g., a carrier). MTS can be described by an AM module set according to their importance to the MTS functions:

$$MTS_{\Psi} = \sum_{j-1}^{1} AM_j + \sum_{j-1+1}^{m} AM_j + \sum_{j-m+1}^{a} AM\,j \qquad (6)$$

3.3 Modular system properties

Unlike the conventional systems, modular systems have the following specific features (**Figure 5**):

3.4 Concepts of flexible technical systems

According to the breakdown in **Figure 2**, the flexible technical systems include modular and structural systems (STS). The difference between these systems is mainly in the autonomy or the sophistication of basic building elements.

The concept of MTS design is to create a complete set of modules (unified units, functional nodes, modular blocks, etc.) and their links in functionally logical (in terms of structure, system, concept, kinematics, etc.) arrangement into a higher functional unit, meeting the required parameters and working functions.

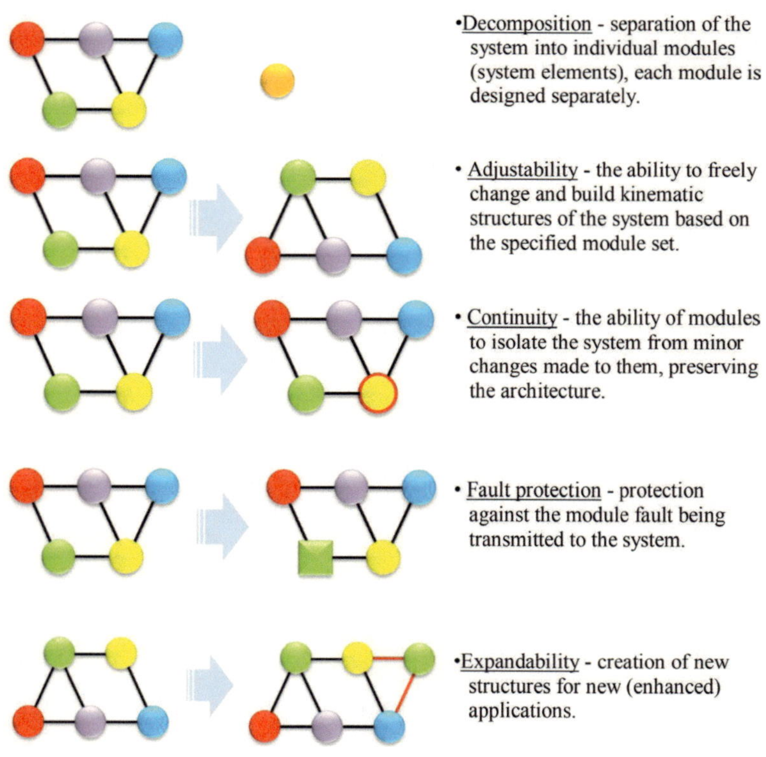

Figure 5.
An illustration of the modular systems' specificities.

Implementing interconnections—arrangement of the assembly of elements and a change of the same:

- *Fixed*—change by external activity, outside the structural system's operation: reconfiguration of kinematic and functional structure of the STS
- *Variables*—change through intrinsic activity in course of the structural system's operation: self-configuration leading to a change in the parameters of the kinematic and functional structure of the STS
- *Fixed*—change through external activity, outside the modular system's operation: reconfiguration of the kinematic and functional MTS structure
- *Variables*—change through intrinsic activity in course of the modular system's operation: self-configuration of the kinematical and the functional MST **structure**

Concept 1—Structural systems with fixed links, such as structural gripper heads by "SCHUNK" (**Figure 6**), STS assembly from a defined number and types of standardized elements (motion units, motion lines, building blocks, unified nodes, etc.), with the possibility of its mechanical conversion outside of operation into new functional and operational STS configurations [7].

Concept 2—Structural systems with variable links, e.g., a turret with multi-spindle heads from Riello Sistemi [8] (**Figure** 7), **STS** assembly from a defined number and types of elements (spindle heads, gripping units, structural blocks, unified nodes, etc.), with the possibility of its mechanical conversion in course of its operation into new functional and operational **STS** configurations.

Concept 3—Reconfigurable modular systems with fixed links, e.g., modules by Riello Sistemi (**Figure 8**), **MTS** assembly from a defined number and types of **AM** autonomous modules (motion units, motion modules, modular blocks, unified nodes, etc.), with the possibility of its mechanical conversion outside operation into new functional and operational **MTS** configurations.

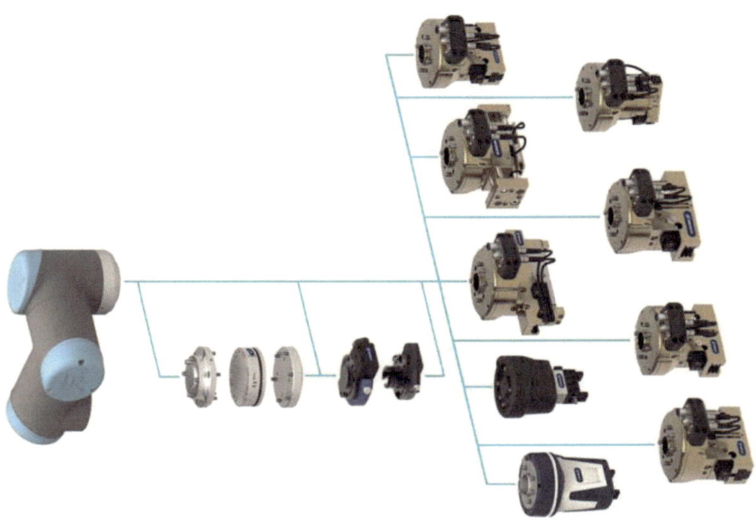

Figure 6.
STS rebuilding concept 1 with fixed links by SCHUNK.

Figure 7.
Concept 2, reconfiguration by adding new elements to the assembly (in production technology, spindle heads) from Riello Sistemi.

Figure 8.
Concept 3, reconfigurable modules by Riello Sistemi [8].

Concept 4—Self-configurable modular systems with variable links, e.g., modules from Riello Sistemi [8] (**Figure 9**) **MTS** assembly from variable number and types of autonomous **AM** modules (motion units, motion modules, modular blocks, unified nodes, etc.), with the possibility of self-conversion in course of the operation into new functional and operational **MTS** configurations.

Figure 9.
Concept 4, self-reconfigurable module sets by Riello Sistemi.

3.5 Modularity of technical systems

A particular MTS architecture made up of AM modules should meet the technical requirements of the application, quality, durability, and safety.

In the MTS system, the AMs are interchangeable—links with other parts of the MTS system are ensured by standard (or special purpose) connectors (interfaces).

AM module features—type and shape of the AM_i module depend on its functionality in the MTS system configuration and parameterization of the resulting requirements (features):

- It can move on top of adjacent modules, or it can rotate or move adjacent modules.
- It may be heterogeneous or homogeneous.
- Depending on the type of positioning and coordination, it can be applied to parallel or serial MTS structure.
- The number of drives and the number of degrees of freedom determine its mobility.
- The type of interface applied determines the capabilities of its metamorphosis.
- An active AM can be built on the rotary or the linear principle of motion.
- Passive AM—connecting AM has no moving parts (the task is to link the active AMs).

The building module architecture is based on the need to appropriately group suitable modules into an MTS architecture that is recurrent for certain types of applications in the form of a structural base.

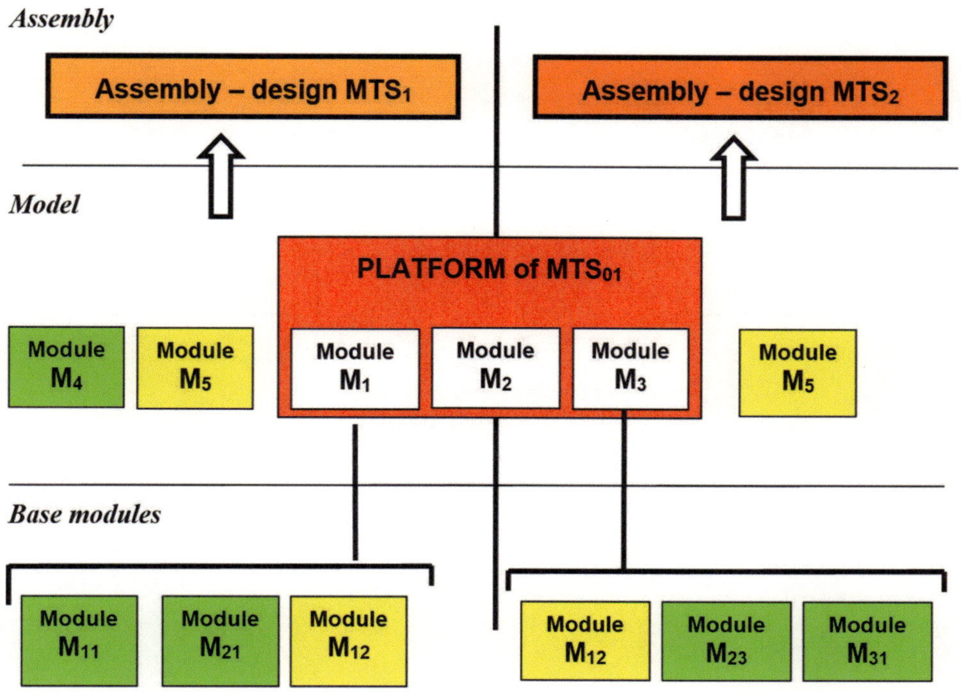

Figure 10.
*Grouping of the modules into a platform in the general **MTS structure**.*

Modules grouped in the MTS architecture (**Figure 10**):

- A *platform* is a set of modules used in multiple complete MTS (modules of platform (MP)) assemblies, e.g., MTS_{01}.
- A set of modules involved in multiple MTS sets (multimachine modules (MM)), e.g., M5.
- A set of modules involved in only one robot assembly (singlemachine modules (MS)), e.g., M4.

The degree of utilization of the unified building modules in the individual MTS design kit expresses the "degree of modularity." In general, the degree of modularity can assume a value of $k_M \in \langle 0, 1 \rangle$ [9].

It is recommended the structure of the MTS assemblies under consideration be compiled into the so-called modular system maps—a clear display of structures of individual assemblies and display of usage of individual building module options in the MTS assemblies.

3.6 Modularity of production technology, features, and characteristics

Assessing modularity—feasible from several points of view. For practical needs of design and operation of production systems, it is appropriate and sufficient to divide modularity into basic groups (in relation to the designed structure) (**Figure 11**).

Functional modularity—linked to the main MTS functions and features (mainly operational). Changing the AM module will change the MTS function (functionality).

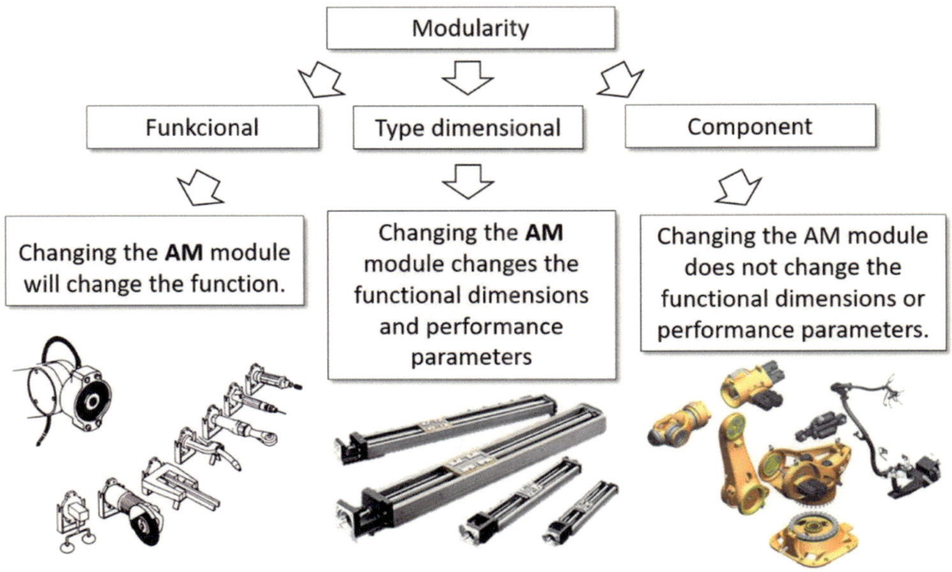

Figure 11.
Modularity breakdown.

Type-dimensional modularity—characterizes the MTS design, its flexible transformation view of another MTS-type series. By changing the AM module, the functional dimensions/performance parameters of the MTS design are changed.

The *AM-type series* (grading of AM parameters or specifying the AM type) is done by distinguishing the parameters as follows:

- Basic groups (typical for structural and functional groups: power, torque, etc.)
- Derived (critical for the user: output, speed, etc.)

Component modularity—characterizes the MTS design in terms of production, maintenance, and service. Changing the AM module does not change the functional dimensions or performance parameters of the MTS design. Such AM module replacement/application makes sense for streamlining the production, service, and maintenance.

AM elements—**components** of one type are physically similar. For this reason, once the conditions are met, the principles of similarity theory [1] can be applied.

Figure 12 shows a dual biaxial modular manipulator designed for varying degrees of load. If the modular assembly is subjected to maximum loads, it is necessary to reduce the motion dynamics to the recommended level or to choose a dimensional range of higher type for the construction of the handling equipment (**Figure 13**) [10].

3.7 Modular production machines

Currently, modular production machines represent advanced machine systems designed on the basis of a mechatronic approach to their design, with the predominant concept of their construction being a three-dimensional and functional modularity structurally built on fixed links.

Conceptually, these structures obey the principles of functional and type and dimension-specific modularity structurally built on fixed links.

Figure 12.
Güdel ZP-type biaxial portal linear motion modules.

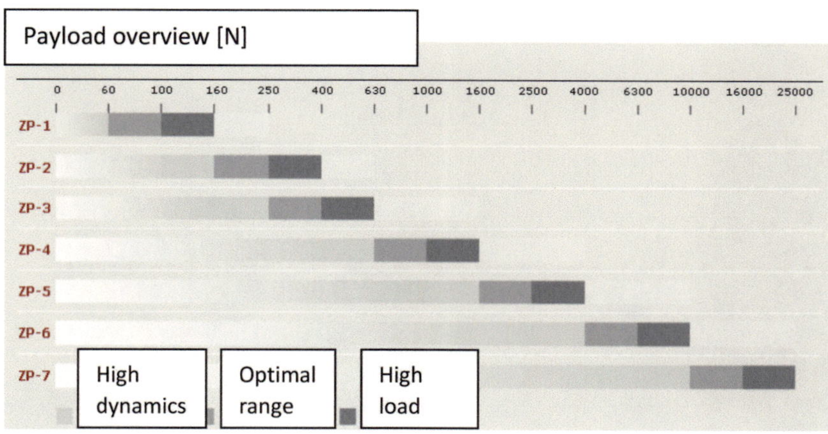

Figure 13.
Güdel type and size-specific linear portal modules of the ZP type.

An example of functional modularity is the IMG industry concept (**Figure 14**). Machining centers, production cells, and transfer lines of varying degrees of complexity can be built from the presented base of modules and platforms.

These two concepts are now strongly prevalent in the development of modular manufacturing technology. A feature of the application of these two concepts in the development of this technique is also their overlap in case of a certain degree of fuzziness of their boundaries or their combination. This direction is particularly pronounced in modular systems enabling the assembly of higher functional units such as machining centers, multiprofessional centers, and production cells.

3.8 Modular robotic systems

Industrial robots and manipulators are implemented as modular systems mainly due to the requirements of flexible automated production systems [4, 11, 12].

From a number of current designs (EPSON, SCHUNK, YAMAHA, KUKA, MOTOMAN, etc.), the solution by SCHUNK will be introduced (**Figure 15**), the

Figure 14.
Modular concept of EMAG production machines [13].

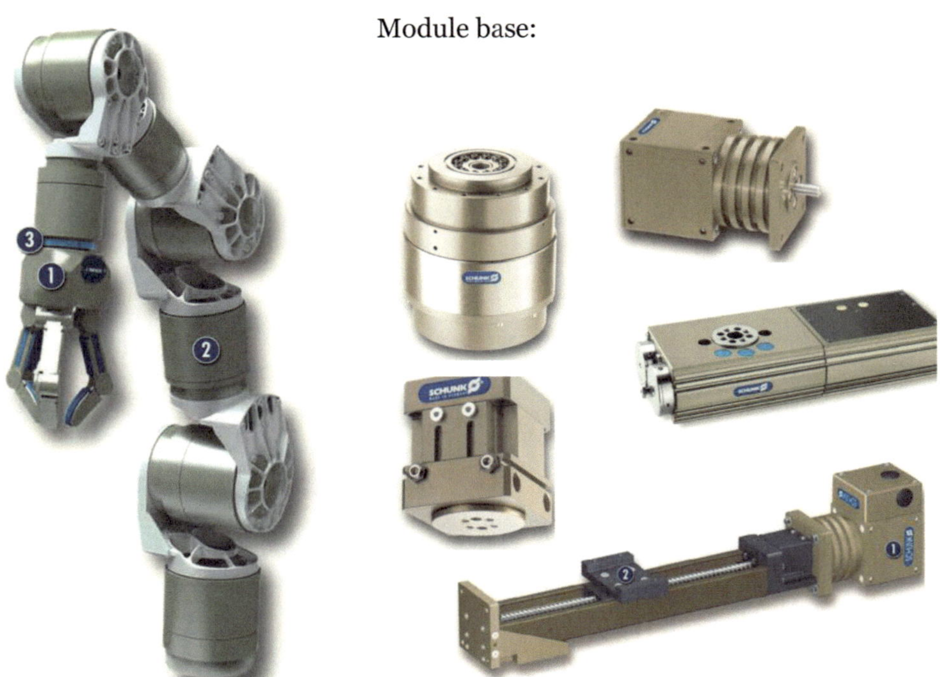

Figure 15.
Modular robot SCHUNK [7]. one- to three-finger gripping effector; 2, rotary module; 3, mechanical interface.

concept of which corresponds to the principles of functional modularity. The presented robot has 7 degrees of freedom when the required number of rotary motion modules can be linked in the series kinematic chain as required. In the design concept, rotary modules are arranged alternately perpendicular to each other, and linked modules are using complex shape interface, which is subject to demanding requirements of stiffness and low weight (a lighter metal material with sufficient strength). This concept is based on the complex structure of the modular system and the links of its elements. The advantage of this design is high flexibility and accommodation of a wide range of requirements of real applications.

4. Design of own universal rotary module with unlimited rotation (URM)

SCHUNK solution (**Figure 15**) can be improved with an innovative custom solution. The presented system of rotary modules called universal rotary module (URM) (**Figure 16**) has any number of degrees of freedom, within the rigidity, load-bearing, and precision characteristics of course. They can be connected to the required number of degrees of freedom (DOF) in the kinematic chain as required. The design concept is changed from the SCHUNK solution so that the rotary

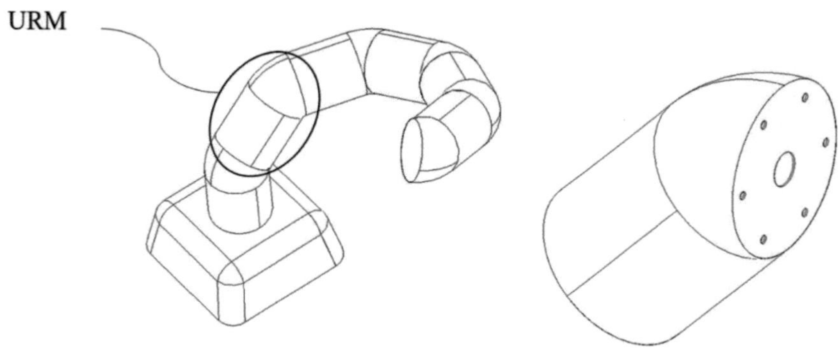

Figure 16.
Basic concept of universal rotational module (URM).

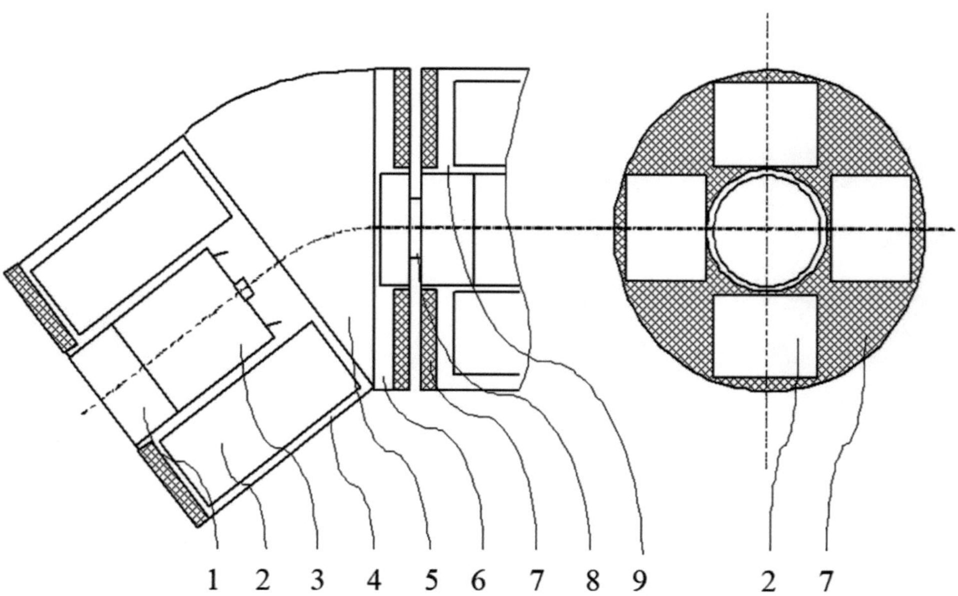

Figure 17.
Basic concept of standalone URM module and section of its structure. 1, reductor; 2, accumulators; 3, servomotor; 4, body; 5, interface; 6, connection panel; 7, coil winding; 8, rotational connection; 9, next URM.

modules are arranged at different angles in the range of 15 to 90° and not perpendicular to each other.

A simple interface is used to connect the modules (**Figure 17**, item 5) which by its curvature determine the extent and reach of the working space of the kinematic structure, which are subject to tough requirements of stiffness and low weight (material of lighter metal with sufficient strength). This concept is based on the complex structure of the modular system and its constraints. The advantage of the solution is high flexibility and coverage of a wide range of requirements of real applications.

Depending on the number of modules involved, a modular manipulator can be created with a working space of different ranges, positions, and shapes (**Figure 18**). The number of modules also determines the number of degrees of freedom of the manipulator. The design and construction of the URM allows the modules themselves to be modified so that their curvature angle may be different from the basic 45°, 90°, and the like. Subsequently, it is possible to assemble modified robot structures. Extension modules can be inserted between the modules to increase the reach of the manipulator arm while maintaining sufficient rigidity of the kinematic chain. All modifications to the mechanical part must be taken into account in the setup and programming of the robot control system.

The main benefits of designing URM-based modular structures are a pair of conveniences:

1. There is no cable bundle inside the modular structure. Neither data nor energy. This means that each module is capable of rotating without limiting the angle of rotation, as is the case of a machine tool spindle. The solution is absolutely wireless. No contact between the moving and the rotating mechanical parts occurs neither is there contact with a brush or carbon brush. Thus, the solution does not generate sparks and is, therefore, suitable to explosive or food handling environment.

2. The URM assembly has its own power accumulators inside each module. Thus, this kinematic structure is resistant to power outages. A certain short house can be operated without power. This is a benefit that other modular solutions do not yet have.

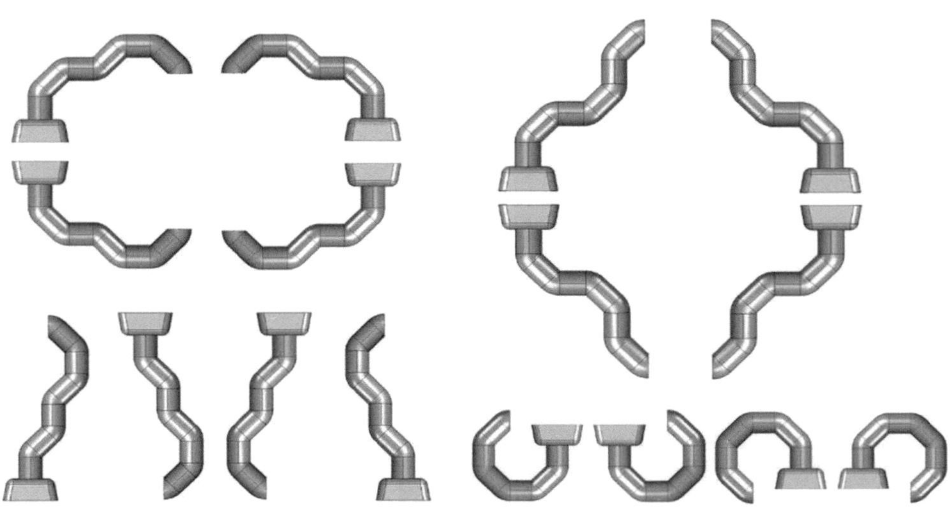

Figure 18.
Examples of reach of a modular assembly made of URM with 6 DOF.

4.1 Design of URM prototype for robotic systems

The URM is developed by one department from Manufacturing Machinery, Faculty of Mechanical Engineering, Technical University of Košice. The result of this solution is a modular system that allows us to assemble modular robots, assembled from identical or type-identical URM01 with unlimited rotational motion. Machines and equipment constructed from these modules are designed to ensure the best possible working range and also to achieve the desired space in the work area (**Figure 19**).

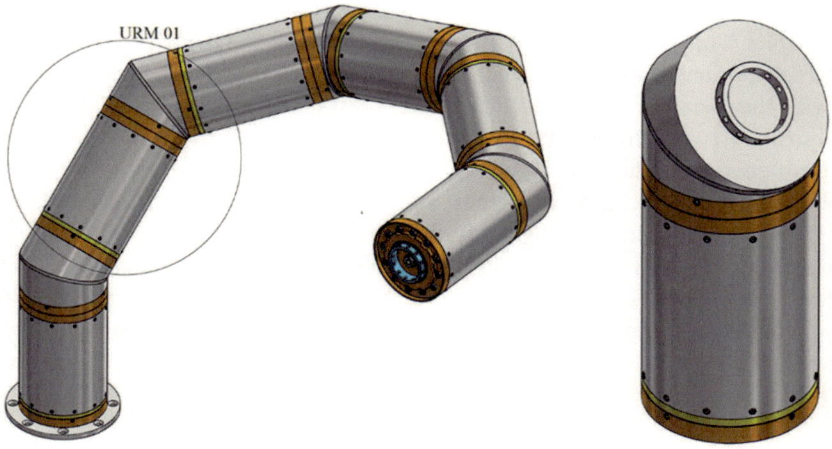

Figure 19.
Manipulator with 6 DOF of movement made of URM 01 modules.

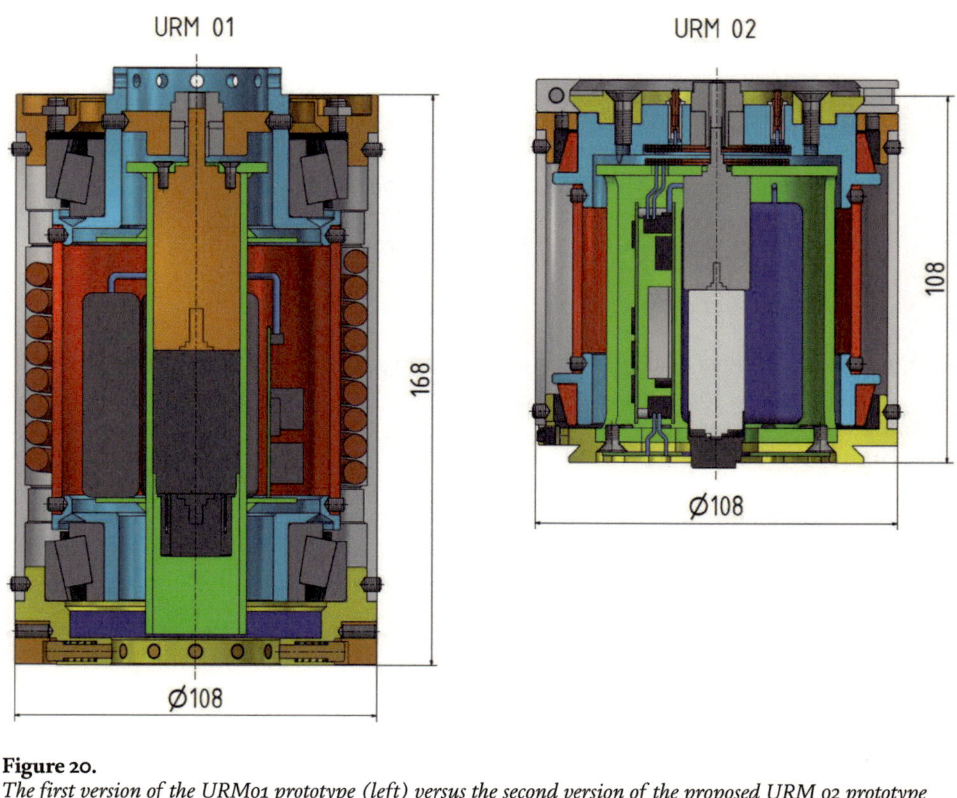

Figure 20.
The first version of the URM01 prototype (left) versus the second version of the proposed URM 02 prototype (right) with basic dimensions.

The main parameter of URM01 is the angle of curvature of the interconnectors. Since this is a homogeneous structure, the curvature of each manipulator module will be the same. A homogeneous structure with 5 DOF of movement was subjected to a series of simulation tests with different angles of curvature of the couplers. In the individual curves of the couplers, structures with interesting shapes are formed. The analysis of the working space of the individual series structures with different angles of curvature of the couplers has brought to light the fact that the more the angle of curvature of the coupler increases, the more the working space of the individual serial structures increases.

The best working space range in all axes has a series structure with 90° curvature of the coupler. This is similar to the standard solutions of SCHUNK, KUKA, etc. In our case, given the curvature of the spacer, it is necessary to consider the possibility of collision with the own modules. This problem can be addressed with the correct control algorithm or software-embedded software limit switches that alert the control system to the limit position of the axis and consequently prevent access beyond that limit.

The prototype URM02, which is conceptually based on the original first variant of URM01, is currently being completed (**Figure 20**). The development of the second-generation URM02 has brought many improvements and possibilities that its predecessor did not contain. Development has taken URM02 to a higher level, making it easier and more efficient to use it in industrial applications. As with any development, the aim was to achieve the best possible results based on the stated objectives and rules of previous research and testing.

5. Conclusions

Benefits of applying modularity to production systems:

- Modules can be developed, manufactured, and tested separately.
- Acceleration (paralleling) of development and production, production, and delivery to the customer.
- Higher component recurrence and simpler logistics.
- Higher level of design variability for the customer.
- Potential expandability.
- Higher flexibility of production volume and assortment (from the manufacturer's point of view).
- Simplification of organization and recycling.
- Acceleration of troubleshooting and service.
- Reduction of development and operating costs.

Disadvantages of applying modularity to production systems:

- Incompatibility with older production systems
- Higher acquisition price compared to conventional systems
- Lower overall efficiency in mass production

Acknowledgements

This work was supported by the Slovak Research and Development Agency under the Contract no. APVV-18-0413 and Research and development of rotary module with an unlimited degree rotation under Contract no. VEGA 1/0437/17.

Appendices and Nomenclature

FMS	flexible manufacturing system
RMS	reconfigurable production system
MTS	modular technical system
MC	machining center
MPC	multiprofessional productions center
MPRC	multiprofessional productions robotic center
AM_i	unified modular unit (functional node, modular block, etc.)
E_{ij}	basic building element of „$_{ij}$ "module
U_i	mutual relations (compatibility of ij module U_i to U_j,or of the ji module U_j to U_i)
X_i	module input parameters (set requirements)
Y_i	module output parameters (properties, operating functions), "a" active, "p" passive
x_{ij}	set of binary interconnections
k_M	degree of modularity
DOF	degrees of freedom

Author details

Jozef Svetlík
Department of Manufacturing Machinery, Faculty of Mechanical Engineering, Technical University of Košice, Košice, Slovakia

*Address all correspondence to: jozef.svetlik@tuke.sk

References

[1] Svetlík J. Modular Architecture of Production Technology. Košice: SjF TU; 2014. pp. 1-141. ISBN: 978-80-553-1928-5

[2] Demeč P, Svetlík J. Modules for Construction of Production Machines and Equipment - Exercise Tutorials. Košice: TU, SjF; 2010. pp. 1-76. ISBN: 978-80-553-0539-4

[3] Marek J et al. Design of CNC Machine Tools. Prague: MM publishing, s.r.o; 2010. pp. 1-421. ISBN: 978-80-254-7980-3

[4] Smrček J. Theory and Construction of Robotic Devices III, Robots on a Modular Principle. Košice: Faculty of Mechanical Engineering, Technical University of Košice; 2012

[5] Warnecke H-J et al. Fractal Plant. Žilina: Slovak Center of Productivity; 2000. pp. 1-208. ISBN: 80-968324-1-7

[6] Smrček J et al. Multiprofessional manufacturing Center with a robot - new approach to creating automated manufacturing systems. Manufacturing Engineering. 2003:32-35. ISSN: 1335-7972

[7] Schunk [Internet]. 2012. Available from: www.schunk.com [Accessed: 06 November 2018]

[8] Riello S. Catalogue [Internet]. 2018. Available from: http://www.riellosistemi.it/en/catalogo/ [Accessed: 12 November 2018]

[9] Gulan L. Modularity as a condition for platform creation. Mechanical Engineering. 2005:38. ISSN: 1335-2938

[10] Güdel [Internet]. 2018. Available from: www.gudel.com [Accessed: 28 January 2018]

[11] Bobovský Z. Design of a robot with a metamorphic kinematic structure [dissertation thesis]. Faculty of Mechanical Engineering, Technical University of Košice; 2009

[12] Skařupa J, Mostýn V. Theory of Industrial Robots. Edition of Scientific and Professional Literature. Vienala, Košice: Faculty of Mechanical Engineering, Technical University of Košice; 2000. ISBN: 80-88922-35-6

[13] EMAG GmbH & Co. KG [Internet]. 2019. Available from: https://www.emag.com/machines/turning-machines/modular-vl/vl-2.html

Chapter 2

Parametric Modeling of Machine Tools

Oleg Krol

Abstract

The chapter deals with the problems of machine tool computer-aided design (CAD) based on the methods and means of parameterization for the main components of metal-cutting machine and equipment in the CAD "APM WinMachine" environment. The models and algorithms of parametric modeling for the configurations of machine tool milling and multioperational type by the criteria of maximum rigidity and minimum reduced load on the front spindle support are developed. The express procedure for generating the transverse layout of the main drive in the multivariate design mode has been implemented.

Keywords: spatial gearbox configuration, transverse layout, multioperational machine, parameterization, 3D model

1. Introduction

Modern computer-aided design (CAD) constructive and technological purposes are widely used methods and means of parameterization to increase productivity, on the one hand, and improve the quality of design decisions made, on the other hand. In the modern "medium-" and "heavy-" class systems, the presence of a parametric model is embedded in the ideology of the CAD systems themselves. The existence of a parametric description of an object is the basis for the entire design process [1].

The process of parametric modeling is associated with the use of model element parameters and the relationships between these parameters, which makes it possible to effectively generate various versions of designed objects using variation of parameters or geometric relations. Unlike traditional 2D and 3D constructions, the use of parameterization tools allows to create a mathematical model of a structure with parameters that, when changed, leads to change the configuration of the structure, relative positions of the parts in the assembly, etc.

One of the designer priorities is to create a conceptual design of the future product and the initial linking of structural elements. Parameterization is a very valuable tool, which allows for a short time to analyze various design schemes and avoid fundamental errors.

Parametric technology corporation (PTC) is considered the pioneer of parameterization, which in 1988 was the first to implement a procedure for creating parametric models.

As is known, the design of machines and tools is primarily associated with the determination of the geometric shapes of detail and their relative position. Therefore, the history of automation is interconnected with the history of computer graphics. Software automation systems had to be invariant with respect to a set of computing tools and equipment input and output of graphic information. This has led to the emergence of various standardization systems. Thus, the standard for the basic graphic system consists of a functional description and specification of graphic functions for various programming languages.

In the field of design automation, the unification of the geometric modeling basic operations gave rise to invariant geometric cores intended for use in various CAD systems. As known, the core includes a library of CAD system basic mathematical functions, which defines and stores 3D forms, processes commands, saves results, and outputs the results of processing. The most widely used are two geometric cores: Parasolid (a product of Unigraphics Solutions) and ACIS (Spatial Technology). The Parasolid core, developed in 1988, becomes the core of solid modeling for CAD/CAM Unigraphics and, since 1996, the industry standard.

Parasolid V19.1 is the first version with support for 64-bit operating systems, using the full power of 64-bit technology to increase the productivity of the creative design process, creating a single 3D modeling platform for the entire product lifecycle management (PLM) industry.

Using the powerful Parasolid core in such well-known CAD Unigraphics allows solving all problems not in external applications, such as in CATIA, where parameterization is performed at the external module level, but at the core level and all its applications working the same way inside the system.

This in-system approach enables the provision of the entire product creation cycle: from the conceptual idea to the implementation within the system itself, without the additional use of external applications. It is also important that providing a unified environment for product development allows you to create a single digital model with which all project participants can work simultaneously. In this case, it is necessary to have sufficiently powerful means of parameterization, allowing for changes of complex structures in large assemblies, to be able to build complex associative links and also certain flexibility, since the product is constantly changing in the design process.

However, in almost all systems, such as Autodesk Mechanical Desktop, Unigraphics, CATIA, I-DEAS, etc., one parameterize of the British company D-CUBED is used. The D-CUBED parameterize includes two components: a sketcher designed to build a parametric profile, on the basis of which a 3D operation will be created, and a mathematical library that allows to link individual parts into assembly structures.

The D-CUBED parameterize, focused on 3D modeling, is ineffective in 2D drawing. The mathematics that successfully works on dozens of profile lines in the 3D system sketcher does not cope with thousands of interrelated elements of drawings. And the need for a complete dimensioning of the D-CUBED parametric model turns the process of parameterization of even a simple drawing into an almost unreal task.

One very useful use of parameterization is the creation of standard element libraries. The cost of creating a parameterization scheme pays off by reusing libraries. So, in the KOMPAS system of the company ASCON [2–5], a simpler approach to the creation of libraries of standard elements has been successfully implemented. It consists in the rejection of the expensive borrowed parameterize and the use of proprietary software for programming a large number of standard elements in plug-in libraries and the transfer to third-party companies of the means for creating such libraries. This allowed the use of an inexpensive software product for obtaining a large set of parameterized libraries.

The process of computer-aided design is constantly being improved. An example of an innovative approach in the field of CAD is the emergence of synchronous modeling (CT) tools proposed by Siemens PLM Software. CAD NX is the flagship CAD/CAM/CAE PLM system (formerly Unigraphics) [6, 7].

This technology allows two main approaches to modeling: parametric design and direct editing. Thanks to synchronous technology, the possibilities of defining the functions of structural elements are expanded, which does not require the description of elements and their limitations manually. New methods for selecting objects automatically recognize logical and intra-element relationships even on models imported from other CAD systems, which contributes to the reuse of design data. For the first time, solutions are proposed for element-wise direct editing without a construction history, which is now available in NX.

CT technology allows setting fixed dimensions and parameters and designing rules at the time of creating or editing a model, without using the history of its creation. Synchronous modeling technology simultaneously synchronizes geometry and design rules by applying a new decision-making mechanism based on an expert system. It allows designers to use geometry from other CAD systems without creating it again. The suggestive selection technology automatically determines the functions of structural elements, without requiring manual description of the elements and their limitations.

Consider the main features of the use of the parameterization toolkit in the APM WinMachine CAD system [8]. In the mode of creating a parametric model, the drawing or any part of it is drawn by creating a special parametric model. The parameters have a numerical expression in drawing units, and their set will determine the dimensional characteristics of the particular part. Parameters are set either numerically or through analytical expressions, and the drawing law is a sequence of drawing commands and their corresponding logical and analytical expressions with the specified parameters. The parametric model created in this way can be inserted into a regular drawing as a parametric unit.

2. Parametric modeling of transverse layout for metal-cutting machine tools

2.1 Build optimal layout

The requirements for the design of specific machine tools vary depending on their types, while the construction of the basic layout is one of the most decisive steps. The layout of the machine is predetermined by the layout of the gearbox and the carrier system of the machine. The design of gearboxes for machine tools (main drive, feed box) is aimed at achieving a large range of rotational speeds (regulation range R_n of modern machines can reach $R_n = 100 \ldots 250$) and high rigidity.

When designing speed boxes (SB), they seek to simplify the design and make it more compact by reducing the number of stages and limiting the gear ratios. So, the reduction of the radial dimensions can be carried out [9–11]:

1. The replacement of the three-shaft box to double-shaft.

2. Rational distribution of gear ratios between several pairs of wheels.

3. Use of parallel transmissions, so that the power is transmitted over the parallel streams and the size of the SB is significantly reduced.

4. Coaxial-mounted shafts, etc.

The use of CAD at various stages of designing assemblies of units and components of the machine involves the integration of a set of design and graphics modules, united by a single design concept with the ability to access common databases [12].

For the whole variety of machines of a certain group (type), it is impossible to use one or two SB structures. Most often, one has to either develop a new design using structural optimization methods or create a new version of the already known prototype design using the parametric optimization method. The parametric model is a sequence of drawing commands with the specified parameters. Parameters are set either numerically or through mathematical expressions.

A feature of the automated design process of a SB is a variety of alternative layout options and the need for adjustments and refinements to the specific features of the design object. The efficiency of the SB design depends on the adopted transverse layout (convolution), including the position of the output shaft. In the existing works on the design of the SB convolutions, the methodology and algorithm for constructing an effective variant of the box design according to the criteria of rigidity and reliability are not given.

In turn, the position of the output shaft in the optimal layout also depends on the position of the resultant cutting force R. Thus, for the range of milling and multioperational machine in the cutting process, tangential P_z and radial P_y components of cutting forces R arise [13–15].

When determining the spatial position of gears that transmit torque to the machine spindle, it is necessary to consider two mutually exclusive situations:

1. Parallelism and unidirectionality of the cutting force R and the resultant force Q in gearing "output shaft spindle" provide the maximum rigidity of the spindle assembly (minimum deflection of the spindle front end). This option is used in machines for finishing processing methods.

2. Parallelism and directivity in opposite directions of the forces R and Q provide the least load on the front support (as the most loaded during the machine operation). In this case, the deflection of the front end of the shaft is maximum, which is permissible only for roughing.

The many options for the SB parts design their mutual arrangement, on the one hand, as well as the need to increase the productivity of the designer, on the other, make it possible to use a parametric modeling apparatus. It is this mechanism that allows reducing the development time of a new or modification of known structures, implemented in all modern CAD systems [15–17].

The parameterization mechanism is characterized by the presence of interconnections and constraints between the geometric objects that make up this structure (as opposed to nonparametric). At the same time, a part of the indicated interrelations and restrictions can be formed automatically when entering graphic information and the rest can be assigned by the user independently.

2.2 Research problem statement

In this chapter, a procedure for constructing parametric models of gearbox transverse layouts for machine tools has been developed. The solved problem is formulated as follows:

To develop such a parametric model of the transverse assembly of the SB, which in one variant will provide the maximum rigidity of the designed machine (its spindle assembly), and in another variant, the minimum reduced load on the front spindle support.

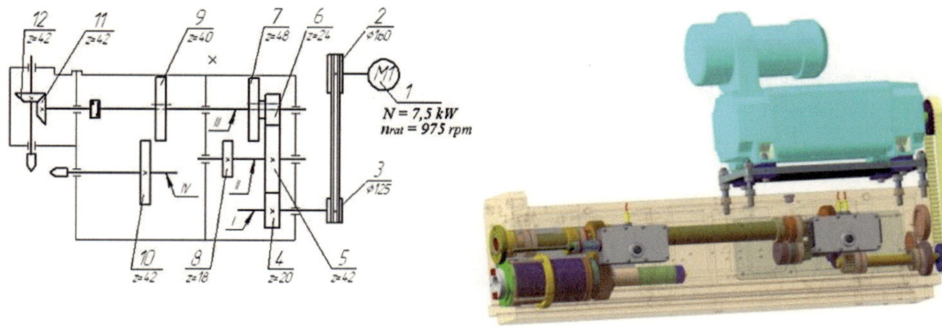

(a) (b)

Figure 1.
Spindle head of the DMB machine with SB: (a) kinematic diagram and (b) 3D model.

2.3 Parametric modeling of the speed box transverse layout for the drilling-milling and boring machine

As a prototype, we choose a horizontal drilling-milling boring machine (DMB machine) with advanced technological capabilities of the model SF68PF4 [18]. The layout of this machine involves moving along the horizontal guides of the headstock (Z axis), to which a vertical head is attached (**Figure 1a**) or additional devices (slotting and angle heads, trunk with a package of disk cutters).

The headstock includes a spindle unit with a tool clamping mechanism and a camshaft transmitting rotation to a horizontal or vertical spindle using an automatic gear shifter. A two-stage gearbox is built into the structure under consideration, as well as a number of other parts and assemblies that ensure the normal operation of the headstock. Rotation from the electric motor through the poly-V-belt is transmitted to the input shaft and through gearing to the camshaft of the gearbox. From the latter, rotation is transmitted to the coupling of the vertical head or to the gear of the horizontal spindle (**Figure 1a**). The 3D model of the DMB machine headstock was developed in the environment of the integrated CAD system KOMPAS-3D (**Figure 1b**) using the specialized application "shafts and mechanical transmissions 3D."

2.4 Optimal layout options

The transverse convolution of the machine during the boring operation, taking into account the location of the cutting forces (P_Z, P_Y) and the forces in the gearing (P_0, P_r), is shown in **Figure 2**.

Analysis of the transverse arrangement shows the nonoptimal spatial position of the machine tool output shaft (n_v) relative to the spindle (n_{sp}), caused by nonparallelism of the resulting forces P and R. This leads to an increase in the reduced load on the front spindle bearing [19] and lowering its carrying capacity. However, this spatial arrangement simplifies the design of the gearbox housing, which is shown in **Figure 3**. In **Figure 3b** presents a three-dimensional model of mechanical gears that implement kinematic connections from an electric motor to a spindle unit. A three-dimensional model of the housing part has also been developed to provide the selected criterion for minimizing the reduced load on the front spindle support.

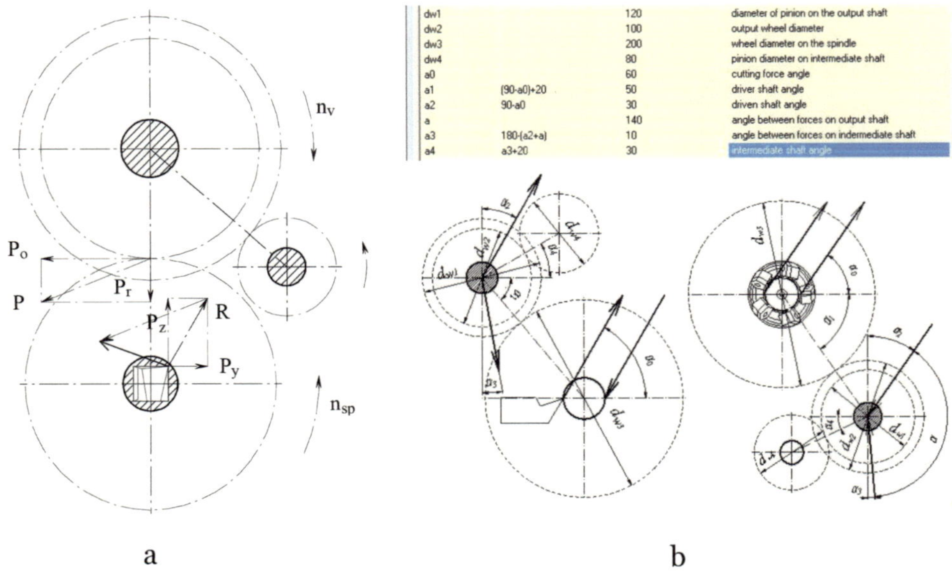

Figure 2.
Transverse layout of the speed box for the machine model SF68VF4: (a) system of forces; and (b) fragment of the optimal layout parametric model.

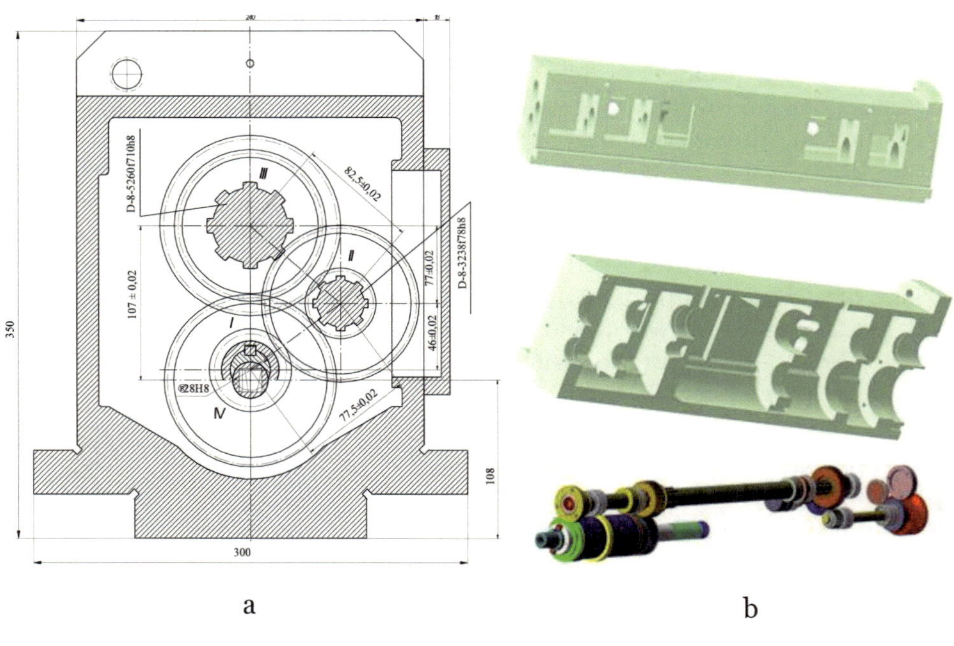

Figure 3.
Speed box of the SF68VF4 machine (transverse layout): (a) design; and (b) housing and kinematics—3D gears.

Rather simple design of the housing differs in two original solutions:

1. On the right side of the housing, there is a joint for mounting the gear in a spatial layout. At the same time, the overall technologically feasible rectangular housing shape with minimal dimensions is maintained.

2. In the lower part of the housing, a spherical shape is deepened, which allows minimizing the size of the housing with fixed initial data.

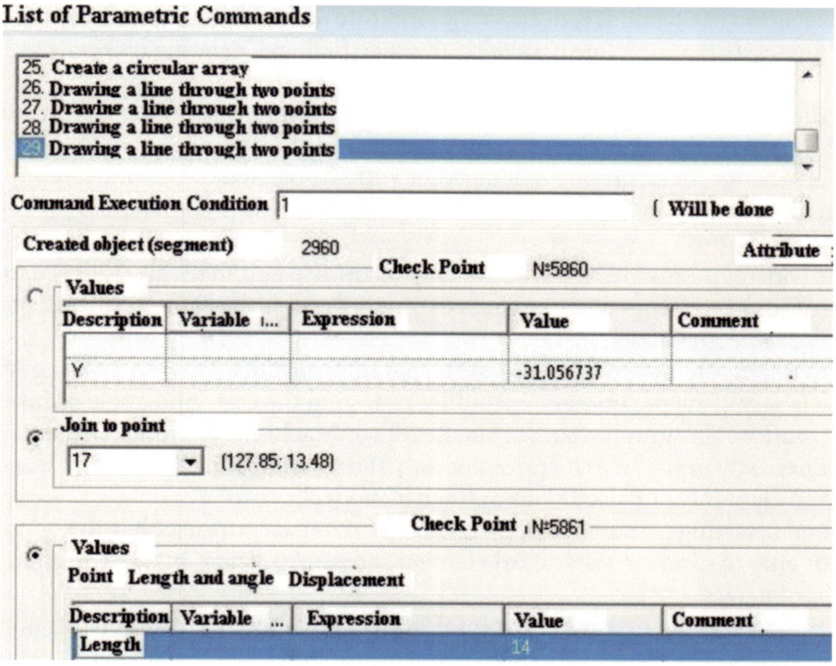

Figure 4.
The command window of the APM graph module.

The presence of the bottom of the nonrectilinear housing is an interesting version for the gearbox of a multifunctional machine. This design allows to implement such a spatial position of the shafts and the spindle, which ensures maximum rigidity of the spindle node and the machine tool as a whole.

At the stage of preliminary design, it is efficient to use a parameterization apparatus, with the help of which it is possible to solve the two-criterion problem of constructing the transverse layout of a horizontal headstock with an integrated two-staged gearbox. For this, we will use the parametric capabilities of the CAD/CAE system of the APM WinMachine system [11, 20].

2.5 Parameterization in the APM graph module

In this system (in the mode of creating a parametric model) is a design of a drawing (or any part of it); a special parametric model is formed. In this case, the APM Graph module is used, where a sequence of drawing commands and their corresponding logical and analytical expressions with the specified parameters are implemented. In **Figure 4** the command window is presented in the task of constructing a transverse layout (e.g., of constructing a "segment through two points").

A parametric model created in this way can be inserted into a regular drawing as a parametric block.

The algorithm for working with the parametric model includes:

- Creation of a parametric model at the command level, which consists in linking graphical objects to those numerical data or communication equations that were set as variables. In this case, auxiliary variables can be used as input in the parameters of subsequent commands.

- Indication of the primitive type being created and its index in commands for creating graphic primitives. Indexes are used in those commands for the execution of which. It is necessary to indicate one or more objects existing in the drawing (drawing a line parallel to the specified one, deleting objects, etc.).

- Editing the list of commands, consisting in the ability to remove a command from the list and replace it with another. Accordingly, the type of the parametric model may change, or errors due that together with the deleted command may appear. In this case any data necessary in subsequent commands was deleted.

The control points created with the object are also automatically indexed and can be used. In subsequent commands to directly access the created control points for the parametric model.

The designer has the ability to insert a parametric block into the drawing with a change in any of its parameters, including changing the scale and angle of inserted block rotation. The graphic part of the APM WinMachine system unified database [20] is similarly organized. It stores not only the numerical parameters of standard parts and elements but also their parametric models.

When inserting a graphical object from the database, the user has the opportunity not only to insert any standard element into the drawing, but also to change any of its parameters.

At the same time, to achieve the minimum load on the front spindle support, **Figure 5b**, it is necessary to adjust the housing design, while the housing connector should be made from the opposite side (**Figure 5a**).

Let us consider the basic design of the control gear for the wide-universal drilling-milling and boring machine with CNC model SF68PF4 [11], built into the two-stage SB of the main movement (**Figure 1**). In the mode of coordination with the rotary table, it provides processing of parts from all sides, as well as coaxial boring of holes without remounting the workpieces. The developed 3D model of

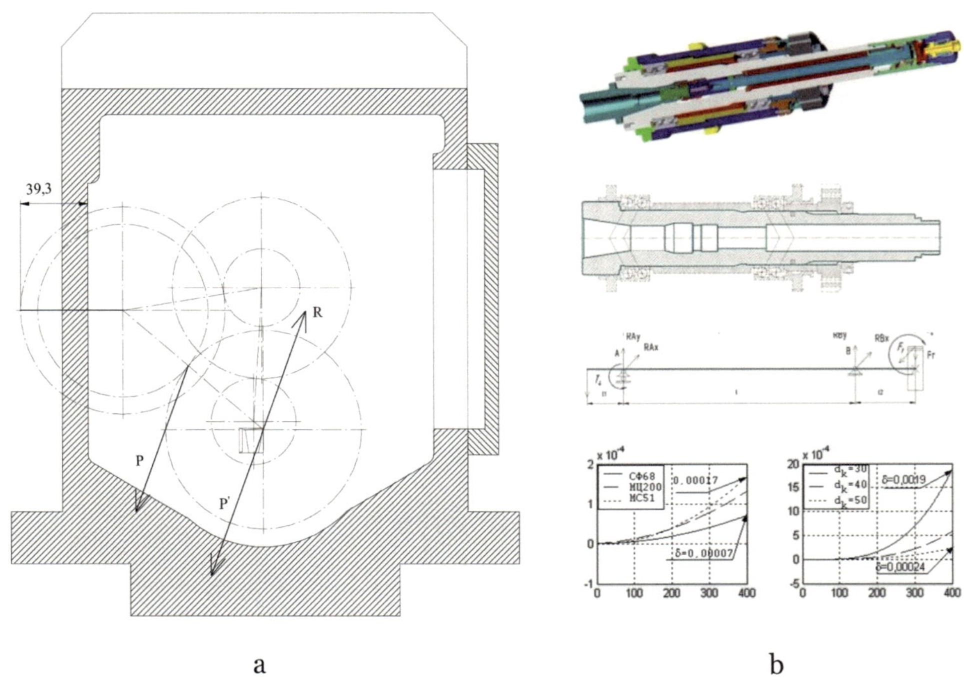

Figure 5.
Transverse layout with the minimum reduced power: (a) design; and (b) study of the spindle stiffness characteristics.

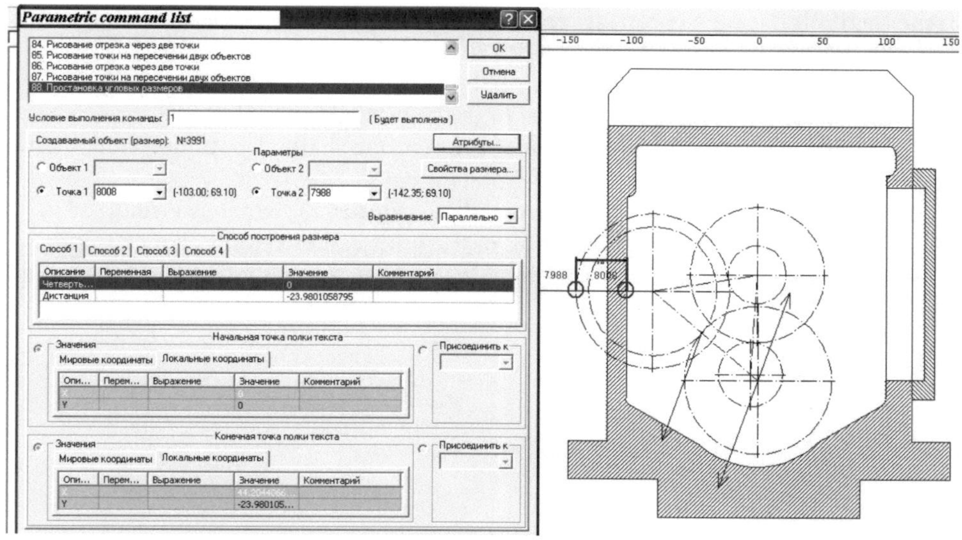

(a) (b)

Figure 6.
Command window of transverse layout parametric modeling: (a) command window; and (b) housing variant with minimum driving force.

the SF68PF4 machine spindle node (SN) using KOMPAS-3D resources for setting parametric connections and associations between the individual components of the 3D assembly is shown in **Figure 5b**.

The SN of this multioperational machine is considered as a beam on two elastic bearings, each of which is mounted on dual angular contact bearings installed according to the "Tandem-O" scheme with preload in the form of different-height bushings. The authors developed a 3D model of SN assembling with a mechanism for automatic tool clamping in the KOMPAS-3D system (**Figure 5b**).

To perform a comprehensive engineering analysis of both individual parts and assemblies, we will use the entire FEM module [19]. This module is equipped with a CAE library that implements solving engineering problems by the finite element method (FEM). In the process of solving, fixations and applied loads are set; matching faces are set (for FEM analysis of the assembly); FEM-mesh generation; calculation and viewing of results in the form of stress and displacement maps are performed.

To study the stiffness, an elastic-deformation model of spindle node two-support construction is built. It takes into account a set of modular equipment (consoles) of various sizes. A feature of the studied object is the presence of two components:

- Unified two-support spindle unit, which can be used in various multioperational machines.
- Tool blocks—a variable component, oriented to a different range of manufactured products.

Using the APM FEM module, all of the above actions were performed, and displacement fields on the set of spindle sections as beams were obtained. The analysis of the compliance characteristics for various multi-operation machines MTs 200PF4, MS51PF3, and SF68PF4 showed that the spindle assembly of the machine SF68PF4 is characterized by minimal flexibility. It has become possible due to the adoption of the optimal spatial layout. A study was made of the change in the flexibility of the SN cantilever part for various standard sizes of tooling: with a spindle cone 30 AT5 in accordance with GOST 15945-82 for the model MS51PF3, with a cone 40 according to GOST 936-82 for model SF68PF4 and a cone 40AT5 according to GOST 15945-82 for model MTs200PF4V (**Figure 5b**).

To assess the change in the position of the gearbox shafts, a parametric modeling program was developed in the APM Graph (**Figure 6**).

Below is an example of a "message variable" (**Figure 7**), which is visualized in the working window in case of violation of the limit values x and k. For example, "Distance between the bottom of the housing and the wheel surface should exceed 24 mm" [21, 22].

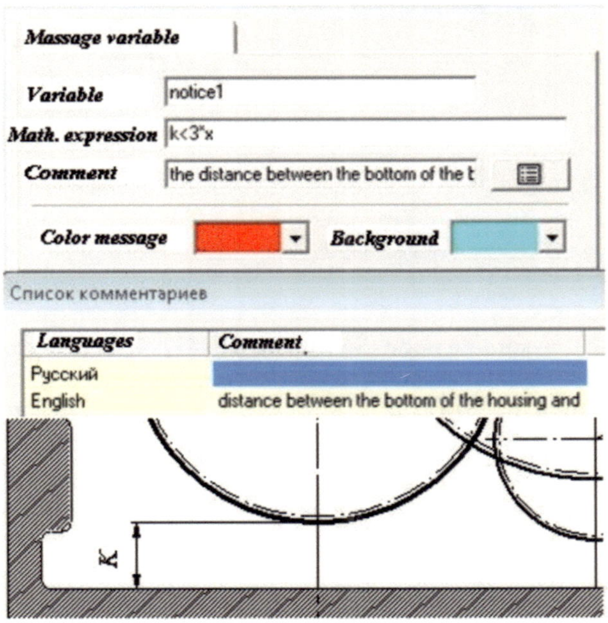

Figure 7.
"Variable-message"—unacceptable distance.

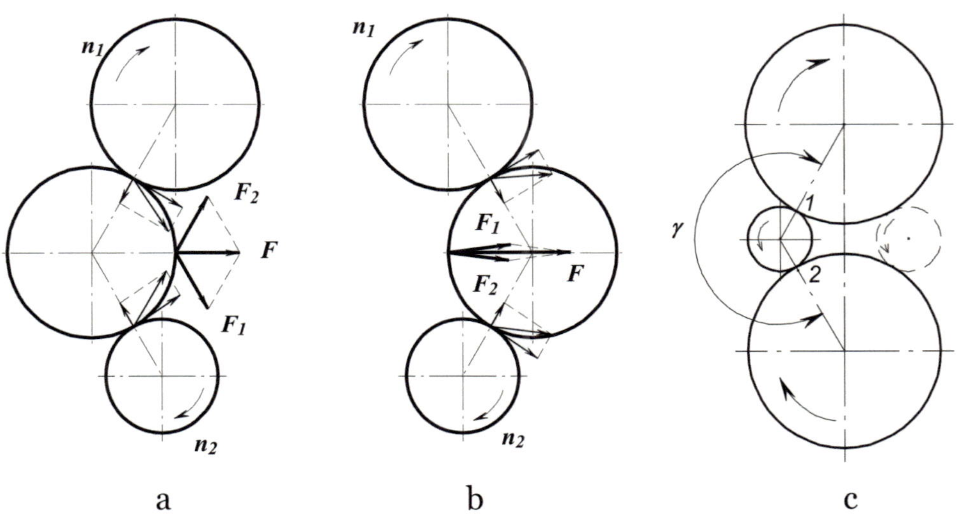

Figure 8.
Schemes of the influence of the intermediate links: (a and b) the intermediate gear; and (c) idle gear.

The arrangement of the wheels relative to each other affects the magnitude acting on the force transmission elements, so that the power characteristics can be improved. So, the location of the intermediate gear wheel (**Figure 8a, b**) affects the forces acting on the shaft bearings of this wheel.

In **Figure 8b**, the forces in the engagement F_1 and F_2 are almost parallel, and the total force F acting on the supports is large. In **Figure 8a**, due to a change in the direction of the forces F_1 and F_2, they are largely compensated, and the resultant forces are less than in **Figure 8b**. It should be noted that for reversing transmission, it is preferable to arrange the axis of the wheels in the same plane.

Rational mounted of wheels [21] also affects the accuracy characteristics of kinematic chains. The error of the idle gear in the kinematic chain (**Figure 8c**) can influence itself in the error of the output link by a magnification doubled. The location of the idle gear for a given direction of rotation also influences. In the diagram (**Figure 8c**), the dashed line shows the best (from the standpoint of accuracy) mounting scheme of the idle gear for a given direction of rotation [23].

The general rule of mounting requires that the transfer of rotation on the idle gear takes place at the minimum angle γ between points of contact 1 and 2. When reversing the transmission of the axis, it is desirable to place on one line, as in the previous review (**Figure 7**).

3. Conclusion

The introduction of parameterization mechanisms in the traditional design process makes the work of the designer as efficient as possible. At the same time, the improvement of the project decision-making process takes place in the following directions:

1. The use of the developed parameterization mechanism significantly increases the efficiency of the metal-cutting machine tool study at various steps of design through the use of parametric models. This approach opens up the possibility of transition to solving complex project problems of a multi-criteria nature.

2. The introduction of the parameterization mechanism contributes to the formulation and solution of designing machine tool problems and their components in the multivariate mode. This significantly increases the level of design decisions made both at the design stage of individual parts and their assembly.

3. A significant effect in increasing designer productivity is the ability to quickly solve the problems of the design process with recurrent information flows (reengineering), when a set of parametric models implements an effective approximation to the best design option.

4. The correctness of the results obtained is due to the use of a wide regulatory base (GOST, departmental normal) in the process of developing parametric models.

5. In the process of parametric modeling, it is possible to introduce into consideration a wide range of parts of a certain class, which reduces the set of necessary models for the design tasks of machine tools and makes more visible the size of the CAD databases of machines.

Based on the proposed methods and parametrization facility, the following results were achieved:

1. Developed parametric models of transverse configurations (layout) of machine tools representative of the turning and milling groups. With the help of these models, built in accordance with the APM WinMachine syntax, it is possible to synthesize optimal transverse layouts both by the criterion of maximum rigidity and the criterion of the minimum load on the front spindle bearing.

2. Using the proposed algorithms for determining the spatial position of nodes in the main motion drive housing, it is possible to determine the distances from the external surfaces of gear wheels to the side walls and the bottom of the housing, as well as to estimate the degree of their approximation to the limit values. This will provide recommendations for reducing the size of the machine drives. On the other hand, the presence of a friendly interface in the APM Graph module promptly provides information to the designer about the unacceptable values of the gaps for rotating parts and the gearbox housing.

3. Based on the developed parametric layout models, recommendations are made for the improvement of housing parts. So, in the design of the main drive housing for the multipurpose machine, it is proposed to change the configuration of the housing bottom in order to ensure optimal spindle stiffness. In the machine drilling, milling and boring group model SF68VF4 proposed to perform the joint of the housing side wall to achieve the optimal design of the spindle unit.

4. Using the integrative capabilities of CAD/CAE/PDM "APM WinMachine" allows the designer to quickly estimate the magnitude of the discrepancy between the optimal and traditional factory solutions. So, with the help of the shaft design module APM Shaft in this work, the difference in the spindle stiffness values for the factory and optimum variants is determined. In this case, the designer receives the calculation form for the main indicators of strength and rigidity.

Author details

Oleg Krol
Volodymyr Dahl East Ukrainian National University, Severodonetsk, Ukraine

*Address all correspondence to: krolos@i.ua

References

[1] Lee K. Basics of CAD (CAD/CAM/CAE). Peter: Saint Petersburg; 2004. p. 560

[2] Fomin EP. Using parametric capabilities of KOMPAS-3D. CAD and Graphics. 2007;**10**:70-74

[3] Krol OS, Sokolov VI. 3D Modeling Of Machine Tools For Designers. Sofia: Prof. Marin Drinov Academy Publishing House of Bulgarian Academy of Sciences; 2018. p. 140. DOI: 10.7546/3d_momtfd.2018

[4] Krol O, Sokolov V. Modeling of carrier system dynamics for metal-cutting machines. In: IEEE Proceedings 2018 International Russian Automation Conference (RusAutoCon); 9-16 September 2018. Sochi: IEEE; 2018. pp. 1-5. DOI: 10.1109/rusautocon.2018.8501799

[5] Krol O, Sokolov V. Modelling of spindle nodes for machining centers. Journal of Physics: Conference Series. 2018;**1084**(012007):1-7. DOI: 10.1088/1742-6596/1084/1/012007

[6] Ushakov D. Synchronous technology—A simulation revolution from siemens PLM software. CAD/CAM/CAE Observer. 2008;**4**(40):1-4

[7] Kurland R. Solid edge with synchronous technology—A revolution in CAD. CAD and Graphics. 2008;**9**:80-83

[8] Rozinsky S, Shanin D, Grigoriev S. Parametric capabilities of the graphic module APM graph of the APM WinMachine system. CAD and Graphics. 2001;**11**:37-40

[9] Bushuev VV. Fundamentals of Designing Machine Tools. Stankin: Moscow; 1992. p. 520

[10] Krol OS, Krol AA. Parametrization of the transverse layout of the main movement drive. Tools reliability and optimization technological systems. Collection of Scientific Papers. 2009;**24**:164-168

[11] Krol OS, Shevchenko SV, Sokolov VI. Design of Metal-Cutting Tools in the Middle of APM WinMachine. Textbook. Lugansk: Publishing house of SNU; 2011. p. 388

[12] Taratynov OV, Averyanov OI, Klepikov VV, et al. Design and Calculation of Metal-Cutting Machines on a Computer: Textbook for Universities. Moscow: Publishing house of MGIU; 2002. p. 384

[13] Krol O, Sokolov V. Parametric modeling of gear cutting tools. In: Advances in Manufacturing II. Lecture Notes in Mechanical Engineering (Manufacturing 2019); 19-22 May 2019. Vol. 4. Poznan, Cham: Springer; 2019. pp. 3-11. DOI: 10.1007/978-3-030-16943-5_1

[14] Krol O, Sokolov V. Parametric modeling of transverse layout for machine tool gearboxes II. In: Lecture Notes in Mechanical Engineering (Manufacturing 2019); 19-22 May 2019. Vol. 4. Poznan, Cham: Springer; 2019. pp. 122-130. DOI: 10.1007/978-3-030-16943-5_11

[15] Krol O, Sokolov V. Parametric Modeling of Machine Tools for Designers. Sofia: Prof. Marin Drinov Academy Publishing House of Bulgarian Academy of Sciences; 2018. p. 112. DOI: 10.7546/PMMTD.2018

[16] Cherpakov BI, Averyanov OI, Adoyan GA, et al. Engineering. Encyclopedia. In: Cherpakov BI, editor. Metal-Cutting Machines and Woodworking Equipment. Vol. 4-7. Moscow: Machinery Engineering Publishing; 2002. p. 672

[17] Avramova TM, Bushuev VV, Gilova LY, et al. In: Bushuev VV, editor. Metal Cutting Machines. Vol. 1. Moscow: Machinery Engineering Publishing; 2012. p. 608

[18] Krol O, Sokolov V. 3D modelling of angular spindle's head for machining centre. Journal of Physics Conference Series. 2019;**1278**(012002):1-9. DOI: 10.1088/1742-6596/1278/1/012002

[19] Shevchenko S, Mukhovaty A, Krol O. Geometric aspects of modifications of tapered roller bearings. Procedia Engineering. 2016;**150**:1107-1112. DOI: 10.1016/j.proeng.2016.07.07.221

[20] Zamriy AA. Practical Training Course CAD/CAE APM WinMachine. Teaching Aid. Moscow: Publishing House of APM; 2007. p. 144

[21] Reshetov DN. Details of Machines. Moscow: Machinery Engineering Publishing; 1989. p. 496

[22] Shevchenko S, Muhovaty A, Krol O. Gear clutch with modified tooth profiles. Procedia Engineering. 2017;**206**:979-984. DOI: 10.1016/j.proeng.2017.10.581

[23] Reshetov DN, Gusenkov AP, Drozdov YN. Mechanical engineering, encyclopedia. In: Reshetov DN, editor. Machine Parts, Structural Strength, Friction, Wear, Lubrication. Vol. IV-1. Moscow: Machinery Engineering Publishing; 1995. p. 864

Chapter 3

Headstock for High Speed Machining - From Machining Analysis to Structural Design

Ľubomír Šooš

Abstract

The progressive technological growth in developed industrial countries is characterized by the increasing range of manufactured parts, the variety of their shapes, and the development and usage of new non-traditional materials. At the same time, demands for high quality and production efficiency must be fulfilled. The essential function of machine tools is to make workpiece surfaces with the required geometry and with the required surface quality under economically efficient conditions. A significant benefit in increasing the efficiency and quality of machined surfaces was the development of high-speed machining. With the application of this machining method, the overall concept of the machine tool and the construction of its individual nodes have changed. The headstock has a significant impact on the quality of the final product and the overall productivity of the machine tool. Machine tools with integrated drive headstocks offer the users much greater performance and reliability. The aim of the presented chapter is the analysis of high-speed machining technology, a description of the structures of high-frequency headstocks and their individual parts, along with the design of a headstock with an integrated drive for the specific case of a machine tool.

Keywords: headstock, high-speed cutting, ball bearings with angular contact, design, testing

1. History, development, and advantages of speed machining

Machining with high cutting speeds is associated with the name Carl Salamon [1]. This German researcher in the 1920s milled, for example, steel with cutting speeds at 440 $m.min^{-1}$, and aluminum at up to 16,500 $m.min^{-1}$. The trials ended in German Patent No 523594 of 1931, creating a series of diagrams describing the impact of the cutting speed on the cutting temperature (**Figure 1**). The experiment focused on machining non-ferrous metals, such as aluminum, copper, and brass, respectively [1]. The theory assumes that at a certain cutting speed (5–10 times higher than in conventional machining), the chip removal temperature at the cutting edge will start to decrease.

His experiments overturned Taylor's theory on "maximum cutting speeds," above which machine damage would occur. Salomon showed that for each tool-work piece couple, there exists a critical speed range at which machining is not

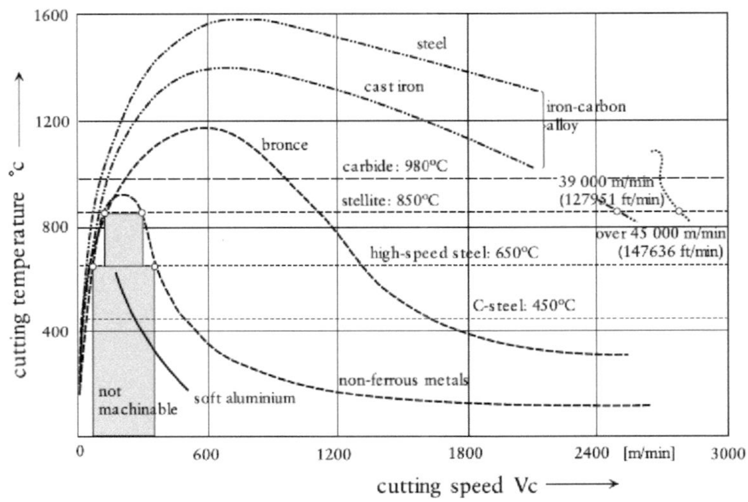

Figure 1.
Machining temperatures at high speeds.

possible. After overcoming this area, we can continue to work, while the temperature of cutting will drop significantly. In the early 1950s, research in the USA carried on from his results. For cutting thin-walled aircraft parts, the Lockhead and Boeing corporations, on a spindle mounted on rolling bearings (n_{max} = 18,000 min^{-1}, P_{elm} = 18 kW), achieved a milling speed of 3000 $m.min^{-1}$, [1–3]. In 1978 in Germany, and a year later in the USA, extensive research focused on the practical usage of high-speed machining was begun. In Germany alone, more than 40 leading firms participated on a project. On the basis of this cooperation at the beginning of 1980 at a university in Darmstadt was constructed an integrated milling spindle unit with asynchronous drive, with a spindle mounted on active magnetic bearings and cutting speeds of 2000–10,000 $m.min^{-1}$ [4]. Over the next 3 years,, an economic variant on a roller bearing was constructed. Partial results from ongoing research confirmed the following advantages of high-speed milling:

- at increased cutting and feed speeds, the cutting force is significantly reduced, [3, 4]. This makes possible the machining of thin-walled parts without special preparations. A drop in cutting forces reduces the demands for rigidity of the whole machine;
- a large part of the heat emerging in the machining process is taken away by chips. This significantly increases the durability of the machine, the work piece remains cold, and the roughness of its surface is decreased with equal or better dimensional precision, which leads to a saving in finishing operations;
- a significant increase in cutting power leads to saving production time so as to save of production costs; and
- high-speed cutting machine provides higher quality surface finish due to the reduced cutting pressure.

A comparison of the advantages and disadvantages of conventional and high-speed machining is shown in the following table (**Table 1**).

Technology of classical machining	High-speed machining technology
The contact time between tool and work is large	Contact time is short
Less accurate work piece	More accurate work piece
Cutting force is large	Cutting force is low
Poor surface finish	Good surface finish
Material removal rate is low	Material removal rate is high
Cutting fluid is required	Cutting fluid is not required

Table 1.
Compare technology of classical machining a HSC technology.

2. Parameters of high-speed machining

Cutting speed values in chip machining are dependent on technology, the material make-up of the cutting machine, and of the machined pieces. Therefore, there exists no unequivocal and general classification of machining according to cutting speeds. In professional literature, we most often encounter the concepts of classical, high-speed, and ultra-high-speed machining. It should be noted, however, that the classifications of the individual authors are considerably different or contradictory. According to individual machining technologies, König is probably the most systematic classification of cutting speeds [3, 4]. This author divides machining into classic and high speed (**Figure 2**). Back in the 1950s, Kronenberg carried out experiments with ultra-high cutting speeds of 9000–720,000 m.min^{-1}. In **Figure 2**, it can be seen that for stretching technology, the area of high-speed machining is in the 30–70 m.min^{-1} range, whereas in this area, cutting speeds from about 5000 to 12,000 m.min^{-1} are used for grinding.

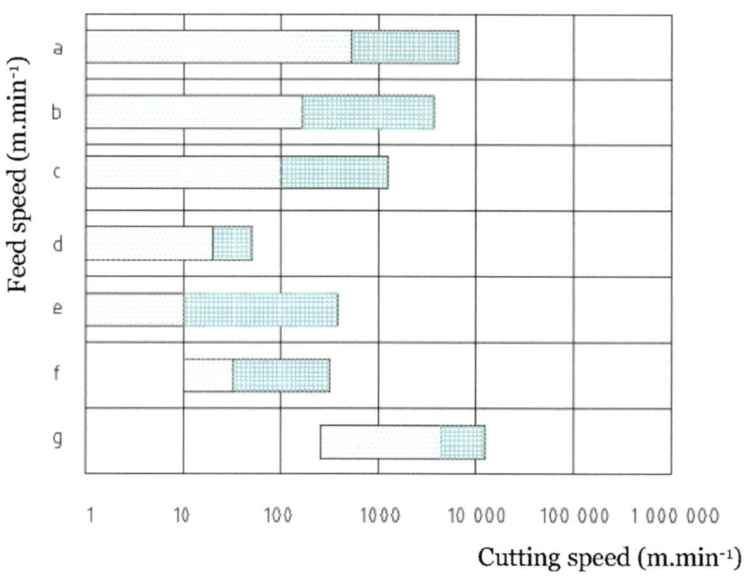

Figure 2.
Cutting speed ranges for chip machining operations. (a) Turning, (b) milling, (c) drilling, (d) stretching, (e) reaming, (f) sawing, (g) grinding, (□) classic machining, and (□) high-speed machining.

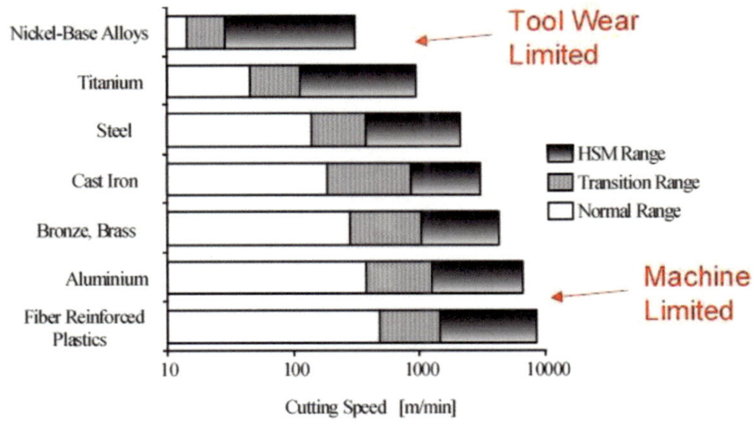

Figure 3.
Cutting speed ranges for milling.

The dependence of cutting speed for individual types of machined material is shown in **Figure 3** [4]. It is clear from the figure that the lowest cutting speed is when milling nickel and its alloys and the highest when milling aluminum and its alloys.

It must be remembered that cutting speed is also a function of the cut material and other accompanying machining conditions (cooling, etc.). To create the most general idea of cutting speed values, we can break down machining according to **Figure 4**.

3. Headstock—heart of machine tool

The issue of high-speed machining is very expansive. It is suitable therefore to divide this area into the conception of the machine as a unit and the development of its individual constructional nodes and elements.

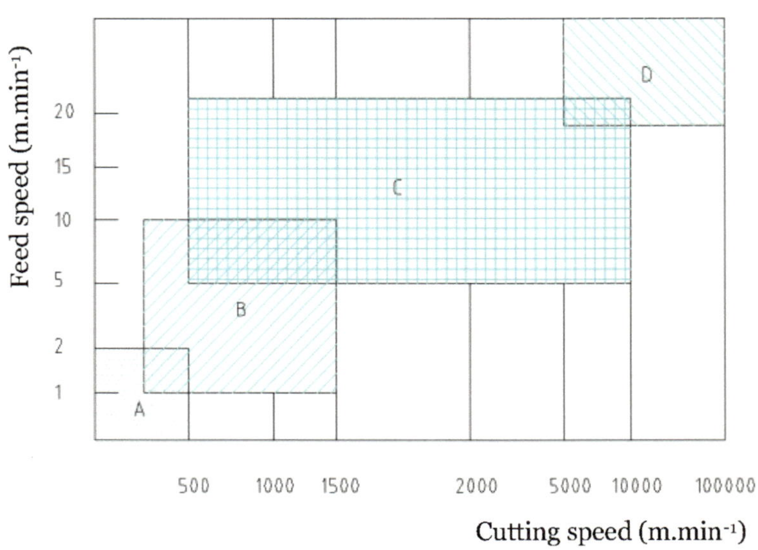

Figure 4.
Outlining of cutting speed ranges for speed machining: (A) classic machining, (B) transitional area, (C) high-speed machining, and (D) ultra-high-speed machining.

Modern machine tools have become more flexible capable of performing a range of programmed tasks.

In the conception of a machine, it is necessary to bring into consideration these factors:

- in the design of a machine's frame, it is important to place emphasis on rigidity and damping capacities. It is very advantageous for this purpose to select the machine frame in high-strength concrete;

- the working area of the machine must be perfectly shroud covered, with good chip transfer and with a suitably selected cooling and control system;

- feed units must be designed with consideration of maximum speeds (vr—15-20 m. min^{-1}), with very short time constants and high strengthening factors (kv ›1.8). This means on one side reducing to a minimum the weight of the moving parts, and on the other, securing maximum rigidity. We can solve this compromise by using high-strength lightweight materials; and

- also important are the aspects of the modularity of the machine construction, together with the rapid replacement of the spindle unit and other machine nodes.

The headstock, as a construction node, has an important position in the overall machine concept. This is for the following reasons:

- the spindle should rotate with the high degree of accuracy. The accuracy of rotation is determined by the axial and radial run out of the spindle nose and these must not exceed certain permissible values that are specified depending upon the required machining accuracy. The rotational accuracy is influenced at the most by the stillness and accuracy of the spindle bearings, particularly the one located at the front end;

- the spindle unit must have high static stiffness. The stiffness of the unit is made up of the stiffness of the spindle unit proper and the spindle bearings. Machining accuracy is influenced on bending, axial as well as torsional stiffness. In series configurations of individual machine nodes, headstock is usually the weakest construction node, which is a limiting member to achieve the required rigidity of the entire machine concept, as a criterion for ensuring the required standstill accuracy;

- the spindle unit must have high dynamic stiffness and damping. Poor dynamic stability of the spindle unit adversely affects the dynamic behavior of the machine tool as a whole; and

- the maximum rotational frequencies of the headstock is a limiting factor of the maximum cutting speed of machine tools and thus of the overall machine production. These maximum rotational frequencies can no longer be ensured for HSC by conventional indirect drives with gear or belt. It should be emphasized that these two factors, stiffness and maximum speed, act in opposite directions.

For high-speed machining, headstocks with integrated drive—"Electrospindles"—are usually used (**Figure 5**) [5]. This has solved the problem of providing rotational frequencies for high cutting speed.

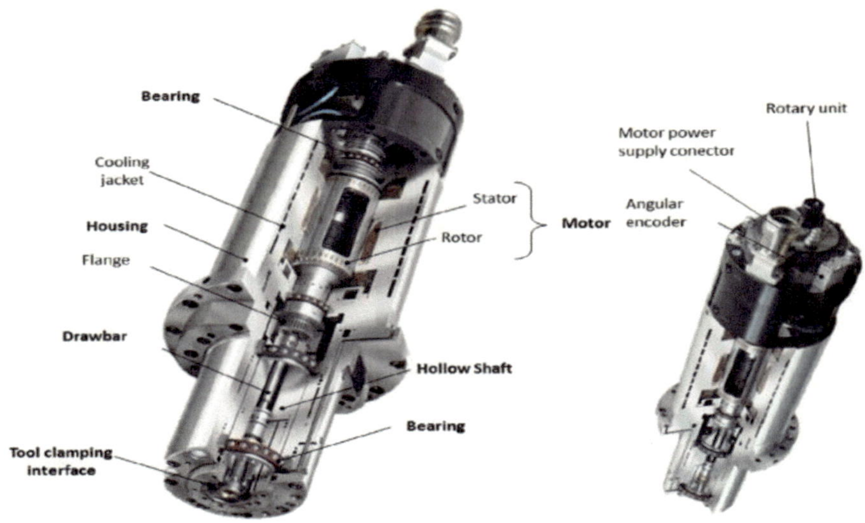

Figure 5.
Spindle unit with integrated drive (SKF) [5].

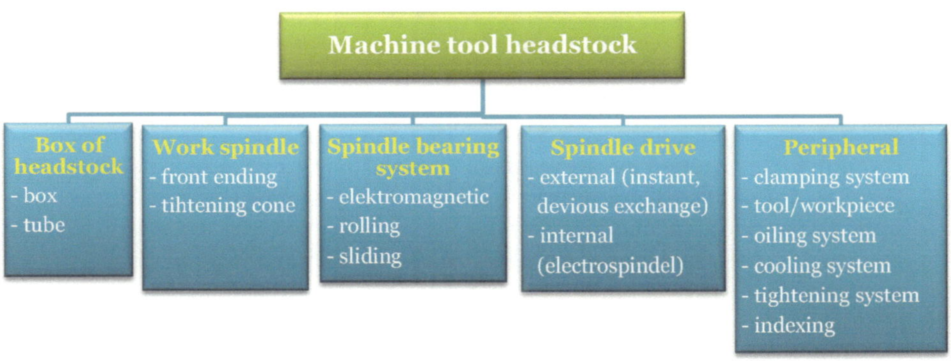

Figure 6.
Headstock morphology.

The electrospindle consists of the particular parts and external peripheries, which together provide the required functions of the whole assembly group (**Figure 6**) [6]. The essential headstock parts include the spindle, bearings system, the tool clamping system or the work piece chucking system, and the body of the headstock. The peripheral devices can include integrated or external systems determined to drive the spindle, lubricate the bearings, provide cooling, spindle indexing, and monitoring.

4. Box of headstock

For high-speed machining, tubular shapes of headstocks like box type are more used. Recently, in addition to tubes made of steel, bodies with a tube wound from fiber composites have recently been used. These include headstocks from Weiss or Step-tech. Based on the elasticity and rigidity knowledge, it is possible to form the approximate solution of every headstock type. Requirements put on the headstock body boxes are:

- maximum symmetry—for the reasons of symmetrical thermal expansions;
- minimum quantity of holes—holes decrease rigidity; and
- statically predestined design—it increases rigidity.

5. Work spindle

The requirements put on the spindle are concentrated on the spindle geometric rigidity, selection of design material, and shape configuration of diameters. The selection of design material for the spindle is conditioned particularly by mechanical properties of the essential core structure, which are by the modulus of elasticity E and by the coefficient of relative damping D. The spindles made of steel comply with the requirements of high static rigidity. The relative spindle quality measure is its specific rigidity, that is, the spindle nose rigidity compared with the spindle weight. The spindle natural frequency and the dynamic characteristics of the headstock are also connected with it. Composite materials (graphite epoxide) start to be used for high-speed spindles. This spindle is lighter, and it does not require such a big diameter [7].

The shape configuration of diameters shall be simple to the maximum possible extent. Those configurations are rational, where the minimum number of graduated diameters can be found and the difference between diameters is determined only by the types and dimensions of applied bearing models.

The spindle end that protrudes from the headstock body is called the front spindle nose. When designing the spindle, the great attention must be paid to the suitable adaptation of the spindle nose so that it can provide the optimum tool clamping (through the clamping shank) or the optimum work piece chucking (e.g., by means of the chuck) [8]. This connection must be a quick, precise, rigid, and reliable one. The type execution and the shape of the spindle nose depend on the technology, type, and size machine tool and on the required accuracy of working.

6. Spindle bearing system

A limiting factor determining cutting speed is bearings. At high frequencies, it must be sufficiently rigid, accurate, and with high durability. The selection of the bearing type in particular supports in the bearing system of the machine tool spindle is always the matter of a compromise among the high rigidity, maximal frequencies of rotation, and offered possibilities of the utilizable building area in the headstock body. In particular, electromagnetic and rolling bearing nodes made of radial angular contact ball bearings are used for receiving spindles for high-speed machining. High revolutions may be achieved by the application of an aero-static bearing whose very low rigidity makes it suitable only for grinding operations.

6.1 Electromagnetic bearings

In the mid-twentieth century, a successful magnetic levitation bearing was successfully demonstrated. This first successful magnetic bearing utilized electromagnets to provide attractive forces in the five degrees of freedom (with rotation being the sixth). Active servo control stabilized the system by using feedback signals from position sensors in each axis of control to vary the currents flowing through the various electromagnets.

Several individual electromagnets, usually from 8 to 12, were arranged in a north-south-north-south configuration around each end of a levitated shaft to provide radial support. This design approach, which results in a multiplicity of magnetic flux reversals around the circumference of the shaft, is known as heteropolar. Most commercially available magnetic bearing systems utilize this technology. A typical heteropolar magnetic bearing system is shown in the below **Figure** 7 [9].

The stator, composed of an array of stationary electromagnets, generates powerful attraction forces that suspend the ferrous rotor shaft in the center of the magnetic field (with the help of an active servo-control unit). The active magnetic bearings are divided into radial, axial, and conical bearings (**Figure 8**).

In addition to the zero mechanical passive resistances, these active bearings have the property that they can determine, for example, the cutting force value, thanks to the active check of the bearing. The reached maximum speed is up to 100,000 min^{-1} and at small special spindles up to 150,000 min^{-1}. The spindle seating on active magnetic bearings uses attractive forces. The spindle position sensors provide the back response for the control system. The sensors send the linear output signal, and they can work in a wide range of operating temperatures. The correct bearing function is ensured by costly control electronics, which prevents faster application of these bearings in the practice. Roller "emergency" bearings are also used in the machine tool spindles carried in the active magnetic bearings (**Figure 9**). The main task of these bearings, which do not work at the normal spindle run, is to provide the trouble-free spindle stop in the case of the sudden electricity blackout.

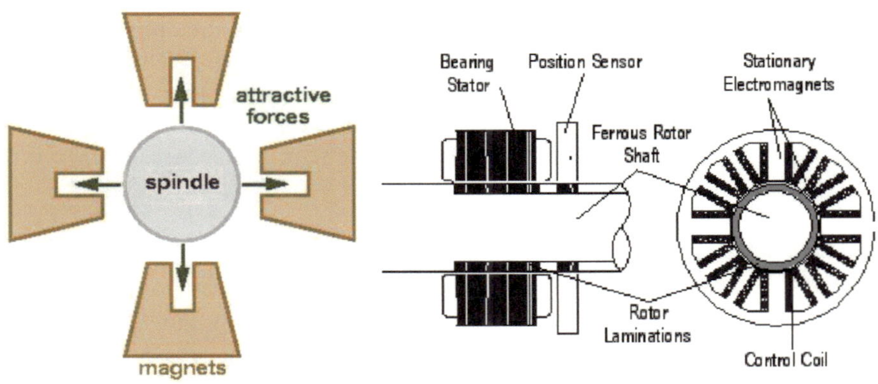

Figure 7.
Principle of electromagnetic bearings.

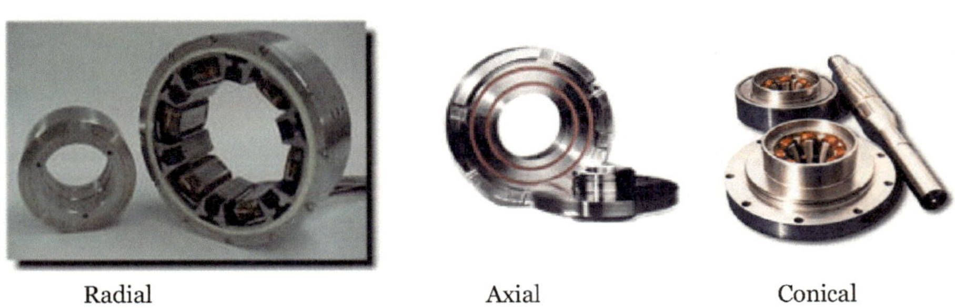

Figure 8.
Type of magnetic bearings.

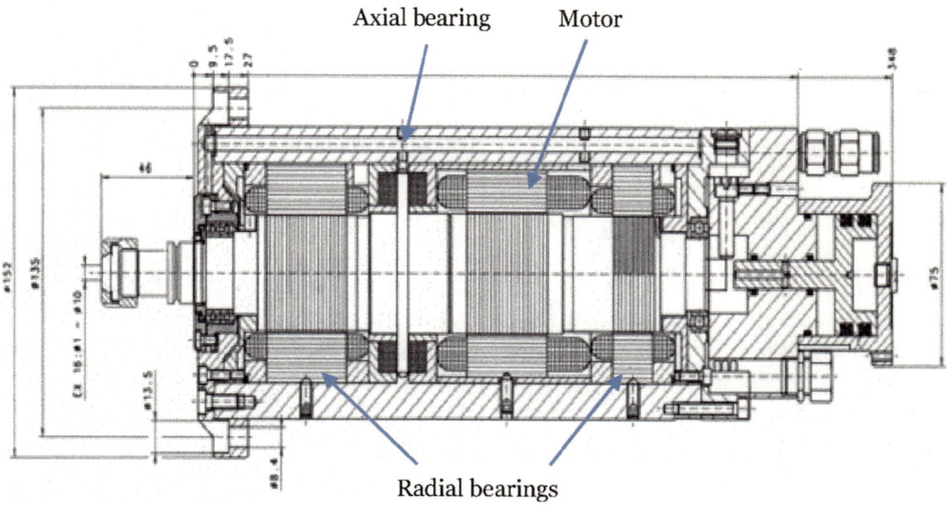

Figure 9.
Electrospindle with electromagnetic bearings (Ibag, HF 120 MA 80 K, n_{max} *= 70,000 min*$^{-1}$*, and P = 11 kW,* M_k *= 1.5 nm), [10].*

6.2 Roller bearings

Radial angular contact ball bearings are used almost exclusively for high-frequency spindle bearings with integrated drive [11]. It is generally valid, that radial ball bearings with angular contact are recently unequivocally the most often used bearings for mounting of high-speed machine tool spindles. The reason is that their different design, their dimensional range, the contact angle values, the preload intensity, and the way of bearing arrangement in the assemblage provide the greatest scope of possibilities how to solve the compromise between the limit speed and maximum rigidity. "Spindle" bearings are manufactured in different dimensional ranges (72, 70, 719, 718) with the design of antifriction body guiding on the inside ring (B) or on the outside ring (A), with different contact angle values (12°, 15°, 25°, and 26°), with the polyamide cage (TB), with the required accuracy (P2, PA9, SP, UP), with various arrangement ways (DB, DF, DT and their combinations), with light (UL), middle (UM), or heavy preload (US) [5]. The bearings made with the higher accuracy degrees are used to seat the spindles. The axial loading capacity of the bearing increases proportionally when the contact angle increases, but the value of limit rotation frequencies decreases. It order to catch bigger radial or axial forces, the bearings are mounted in assemblages created from three, four, or five bearings. Radial load is distributed to all bearings in the group (shape arrangement), and axial load is distributed to all bearings joined behind each other (direction arrangement).

6.2.1 Observed parameters of the bearing groups

The important parameters of the bearing groups specified to seat the working spindles at machine tools are:

- run accuracy;
- durability;

- rigidity;
- high-speed run; and
- temperature.

6.2.1.1 Run accuracy

The run accuracy spindle bearing system is limited by the accuracy of bearings and by the accuracy of bearing surfaces—connection parts. The accuracy of antifriction bearings is understood as the accuracy of their dimensions and run. The limit values for the accuracy of dimensions and run are mentioned in ISO 492 and ISO 199 standards. The accuracy of connection parts is understood as geometric shape and position deviations which can be admissible at the manufacture of the spindle and headstock box. The bearing manufacturer prescribes the admissible geometric shape and position deviations of bearing surfaces (**Figure 10**). At the assembly of bearing, it is necessary to observe matching of inside and outside bearing diameters to provide the required radial preload.

6.2.1.2 Durability

The calculation of bearing durability is generally known [6]. It is described by the international ISO 281/l standard. When durability is calculated, we usually use the modified equation of durability that expresses the durability in operation hours. The following relation is used for the bearing durability in hours:

$$L_{h10} = \left(\frac{C_d}{P}\right)^p \cdot \frac{10^6}{60.n_s} \ [\mathrm{h}] \qquad (1)$$

where P is the equivalent dynamic load [N];
C_d is the dynamic loading capacity of the bearing [N];
exponent: $p = 3$, for ball bearings;
$p = 10/3$, for needle, spherical-roller and tapered roller bearings; and
n_s is mean frequencies of bearing rotation $[\mathrm{min}^{-1}]$.

The equivalent dynamic load P at roller bearings corresponds to the intensity of reactions in the particular supports. However, the methodology is not unified how to calculate the equivalent load at bearing groups made of the radial angular contact ball bearings.

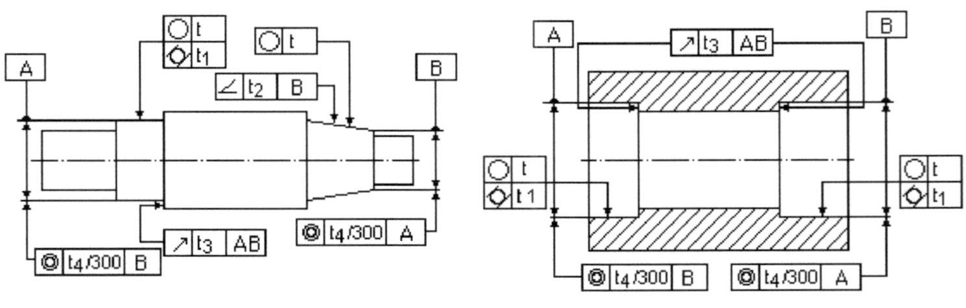

Figure 10.
Prescribed shape and position deviations (SKF).

The spindle bearings transfer the combined radial-axial load. When the selected bearing type (selected bearings) is calculated, the combined radial-axial load is recalculated to the so-called equivalent dynamic load:

$$P = X.F_r + YF_a\ [\text{N}] \tag{2}$$

where
F_r is the radial force [N];
X is the radial coefficient;
F_a is the axial force [N]; and
Y is the axial coefficient.

6.2.1.3 Rigidity

The significance of the bearing rigidity in the particular supports is considerable at the spindles having a bigger diameter, where the rigidity of the bearing assemblage in the particular supports is the limiting factor necessary to reach the required rigidity of the complete seating, as the tool how to provide its accurate operation. The total rigidity is the criterion of the body resistance against the influence of external forces.

The rigidity of the bearing assemblage made from the radial angular contact ball bearings can be described mathematically as the multiple parametric function [11].

$$C_{r,a} = f\ \left(i, z, d_w, \alpha, F_p, \left(\delta_{ps}\right), F_{r,a}, F_n, T\right) \tag{3}$$

It depends on the number of bearings i, dimensional rank, size and design of bearings z, d_w, contact angle α, preload size F_p, or deformations due to preload δ_{ps}, and frame conditions (bearing accuracy, assembly, and cooling).

Three essential states can generally take place in the bearing assemblage made from the radial ball bearings [6]:

- the preload state [e.g., the assemblage joined from two shape-arranged bearings (**Figure 11a**)];
- the preload axially loaded state [the TBT assemblage loaded by the axial force (**Figure 11b**)]; and

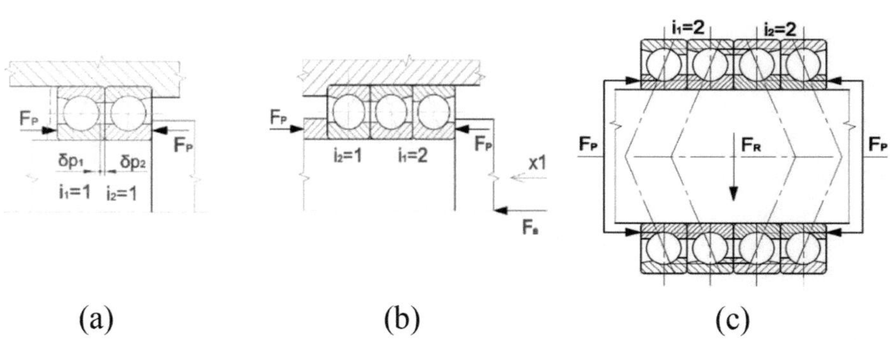

Figure 11.
Essential states of bearing assemblages: (a) DB preload state, (b) TBT preload—axially loaded state, and (c) QBC preload -radially loaded state.

- the preload radially loaded state [the QBC assemblage loaded by the radial force (**Figure 11c**)].

Preload of the spindle bearings at the spindle assembly enables to increase the working accuracy and rigidity of the whole seating. On the other hand, the increased preload initiates the temperature origination in the bearing, which has the negative influence on critical rotation frequencies of the bearing or of the bearing assemblage. Two angular contact bearings are preload by the force F_p according to ČSN/STN 024615,

$$F_p = k.C_d.10^{-2} \tag{4}$$

The preload value for more bearings will be increased depending on the relation:

$$F_{ps} = k.C_d.i^{0,7}.10^{-2} \tag{5}$$

6.2.1.4 Axial rigidity

The axial rigidity importance comes to the foreground especially at facing, milling, drilling, and grinding. In the system "spindle-bearing," the axial forces are almost always caught by the point-contact bearings. The axial rigidity is then given by the relation:

$$C_a = \frac{F_a}{\delta_a} \tag{6}$$

The following is valid for the approximate axial rigidity value according to [11]:

$$C_{az} = \frac{3.10^{-3}}{2} z^{\frac{2}{3}} \cdot k_\delta^{\frac{2}{3}} \cdot i_1^{\frac{2}{3}} \cdot F_p^{\frac{1}{3}} \cdot \sin^{\frac{5}{3}}\alpha_1 \left[1 + \frac{i_2^{\frac{2}{3}} \cdot \sin^{\frac{5}{3}}\alpha_1}{i_1^{\frac{2}{3}} \cdot \sin^{\frac{5}{3}}\alpha_2}\right] \tag{7}$$

After the omission of the contact angle change due to the axial force and under the presumption that the contact angles are the same ones at both joined groups, the relation becomes the simplified form:

$$C_{az} = \frac{3.10^{-3}}{2} z^{\frac{2}{3}} \cdot k_\delta^{\frac{2}{3}} \cdot i_1^{\frac{2}{3}} \cdot F_p^{\frac{1}{3}} \cdot \sin^{\frac{5}{3}}\alpha_1 \left[1 + \frac{i_2^{\frac{2}{3}}}{i_1^{\frac{2}{3}}}\right] \tag{8}$$

6.2.1.5 Radial rigidity

$$C_r = \frac{F_r}{\delta_r} \tag{9}$$

For the reason that the load is not distributed equally, the rigidity calculation is rather difficult and it cannot be almost realized without application of computer technology. It is necessary to determine theoretically and to verify experimentally the deformation course on the load at the preload point-contact bearing groups. The research of the bearing groups made from the radial angular contact ball bearings [12] showed that the deformation course is almost linear at the preload bearing groups up to the certain critical load. For the calculation and testing of radial ball bearings arrangement to nodes, we have developed an expert mathematical model allowing calculation of stiffness, limit frequencies, and bearing node durability.

Based on this knowledge, the simplified equations for the calculation of the mean rigidity value were deduced in works [12, 13].

6.2.1.6 Directional rigidity

$$C_{\mathrm{rsi}} = \frac{3.10^{-3}}{4} . z^{2/3} . k_{\delta}^{2/3} \cdot i^{2/3} \cdot \mathrm{F}_{p}^{1/3} \cdot \frac{\cos^{2}(\alpha)}{\sin^{1/3}(\alpha)} \tag{10}$$

The resulting radial rigidity of the bearing group with the shape-arranged bearings will then be:

$$C_{\mathrm{rz}} = C_{\mathrm{rs1}} + C_{rs2} \tag{11}$$

The following relation was deduced according to [13] for the approximate radial rigidity value of the bearing assemblage made from two shape-arranged groups:

$$C_{\mathrm{rz}} = \frac{3.10^{-3}}{4} \cdot z^{2/3} \cdot k_{\delta}^{2/3} \cdot i_{1}^{2/3} \cdot \mathrm{F}_{p}^{1/3} \cdot \frac{\cos^{2}(\alpha_{1})}{\sin^{1/3}(\alpha_{1})} \cdot \left[1 + \frac{i_{2}^{2/3} \cdot \cos^{2}(\alpha_{2}) \cdot \sin^{1/3}(\alpha_{1})}{i_{1}^{2/3} \cdot \cos^{2}(\alpha_{1}) \cdot \sin^{1/3}(\alpha_{2})}\right] \tag{12}$$

At the omission of the contact angle change due to the axial force and under the presumption that the contact angles are the same ones at both joined groups, the relation becomes the simplified form:

$$C_{\mathrm{rz}} = \frac{3.10^{-3}}{4} \cdot z^{2/3} \cdot k_{\delta}^{2/3} \cdot i_{1}^{2/3} \cdot \mathrm{F}_{p}^{1/3} \cdot \frac{\cos^{2}(\alpha_{1})}{\sin^{1/3}(\alpha_{1})} \cdot \left[1 + \frac{i_{2}^{2/3}}{i_{1}^{2/3}}\right] \tag{13}$$

where k_{δ} we can calculate according to the equation

$$k_{\delta} = \sqrt{1,25 d_{w}} \tag{14}$$

Under the presumption that the contact angles $\alpha_1 = \alpha_2$ **are the same ones at the shape-arranged bearings in the group or** $i_2 = 0$ for the direction-arranged bearings in the group, the relationship between radial and axial stiffness is simplified as:

$$C_{rz} = \frac{C_{az}}{2} \cdot \frac{1}{tg2\alpha} \tag{15}$$

6.2.1.7 High-speed run

The high-speed run criterion is the quality criterion of the node regarding to the reached frequencies of rotation. Regarding to the high-speed run of the bearing nodes, the node systems are analyzed in work [6]. The particular designing solutions of the existing seating are divided into three essential groups in this work. The high-speed run parameter can reach the value K = (2–2.7).106 mm. min^{-1} at the special high frequency groups. For the limit values, it is suitable to use the special bearings with the optimized design, high accuracy, and with the utilization of materials having the favorable physical and mechanical properties (e.g., silicon nitride Si_3N_4).

$$K_r = n_z \cdot d_s \left[\min{}^{-1} \cdot \text{mm}\right] \tag{16}$$

The following relation is used for the determination of the critical frequencies of the bearing groups n_z.

$$n_z = n_{l\,\max} \cdot f_1 \cdot f_2 \cdot f_3 \cdot f_4 \dots \dots f_n \left[\min{}^{-1}\right] \tag{17}$$

where $n_{l\,\max}$ is the critical frequencies of the bearing rotation and f_i is the coefficient describing the bearing group and conditions of its work (the number of bearings, preload, accuracy of bearings, kinematics, heat removal, lubrication, etc.). Their importance is different in dependence of the particular sources.

The reduction of antifriction body dimensions results in the decrease of the centrifugal force, for which the following is valid:

$$F_o = m \frac{d_s}{2} \cdot \omega^2 \tag{18}$$

where m is the antifriction body weight and d_s is the bearing mean diameter.

Such bearings are economical and reliable. The issue of decrease in centrifugal forces at high-frequency rotations is solved by reducing the weight of the rolling elements. This is achieved by changing the dimensional series of bearings and by changing the ball material.

Using bearings with smaller cross-sections, for example, 718, 719 instead of bearings with bigger cross-sections (70, 72), reduces the diameter of the balls. Reducing the diameter of the rolling elements makes it possible to increase the high frequency of rotation of the bearing (**Figure 12a**) and at the same time increases the number of balls to achieve higher bearing stiffness (**Figure 12b**) [5]. With constant external diameters, the internal diameter of the bearings increases, which is suitable from the standpoint of reducing spindle deflection, increasing its drilling, and increasing the critical revolutions of the spindle.

Roller bearings as well as ball bearings can be made as so-called "hybrid ones," which means that the bearing rings are made of steel and antifriction bodies are ceramic. The advantage of hybrid bearings by the same size compared to steel bearings is their lower centrifugal forces, frictional moment, and higher radial and axial stiffness (**Figure 13**). Disadvantages include the high manufacturing costs

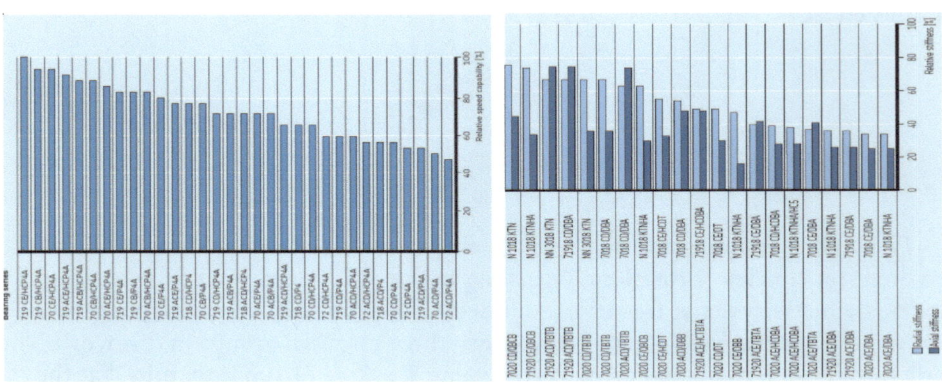

(a) (b)

Figure 12.
Changing the dimensional series of bearings—contact ball bearings (SKF) [5] (a) relative speed capability, and (b) relative stiffness.

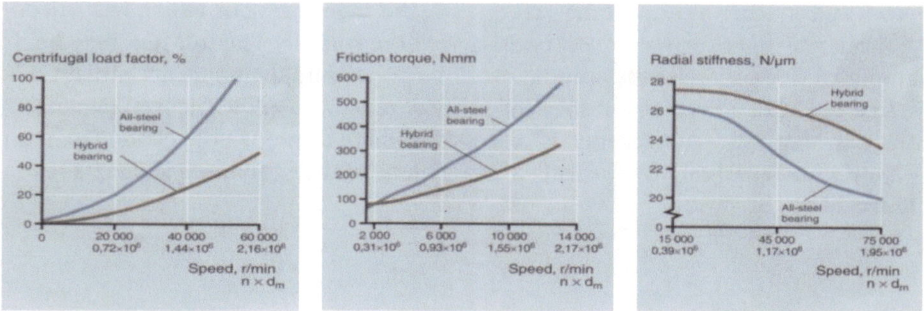

Centrifugal load Friction torque Radial stiffness

Figure 13.
Hybrid bearings (SKF).

(up to 10 times) of rolling elements, still persisting problems with the homogeneity of ceramic materials, and the identification of failures.

6.2.1.8 Temperature

In the bearing groups, where no external heat sources act, the shaft temperature, the spindle temperature as well as the temperature of inside bearing rings and of the antifriction bodies are higher than the temperature of external bearing rings and of the headstock body sleeve. Due to the heat drop at the same expansibility coefficient, the dilatation of the spindle, bearing rings, and balls is bigger than the expansibility of the surrounding parts in the radial direction as well as in the axial direction. According to [11], its value is described by the equation:

$$\Delta l_{r,a} = \upsilon_t \cdot l_{r,a} \cdot \Delta t \tag{19}$$

For headstocks where high demands are placed on the range of rotational frequencies or temperature, it is advantageous to vary the amount of bearing preload directly during work. In order to increase the speed ranges and the service life of the spindle bearing, due consideration must be given to temperature optimization of the bearing when designing the spindle. The temperature of the bearing system varies depending on the temperature gradient, type and arrangement of the bearings (DB, DF, DT), assemblies, contact angle, bearing size, and the distances of bearings in the note and of the individual supports. There are known systems of active control of bearing preload of high-frequency headstocks and peripheral

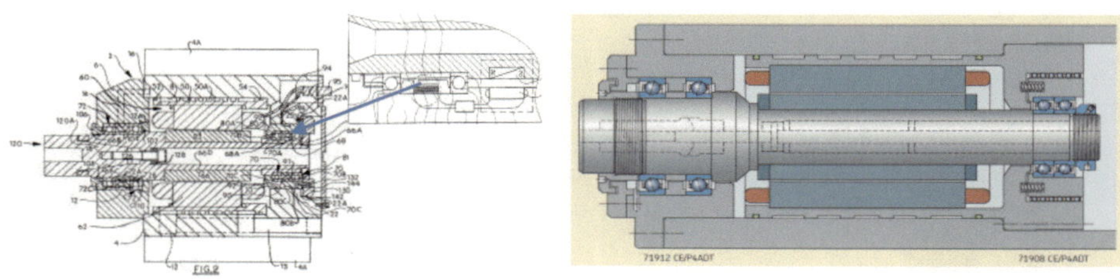

(a) (b)

Figure 14.
Temperature change compensation. (a) US06422757; Active piezoelectric spindle bearings preload adjustment mechanism, *[14]. (b) Motor spindle SKF with movable rear support [5].*

devices and sensors of important parameters, the monitoring of which has a decisive influence on ensuring correct operation of the spindle. The solution may be, for example, active piezoelectric spindle bearings preload adjustment mechanism (**Figure 14a**) [14]. Bearings in the rear support must also allow thermal expansion of the entire spindle. Advantageously, it is possible to minimize the change in bias in the bearing by resolving the bearing arrangement in the individual supports (**Figure 14b**) [5].

7. Spindle motor

Desired performance and revolution characteristics place ever increasing demands on the construction of the spindle unit. The type of propulsion and bearing is the decisive component for providing the stated characteristics. Incorporating the drive directly into the spindle unit has successfully solved transmission problems at high speeds. In this way, the stress from the drive forces onto the spindle is eliminated and its accuracy is increased (**Figure 15**).

Both single-direction and alternating drives can in principle be used for integrated spindle units. Despite very good control properties, DC drives have known operational and technical drawbacks resulting from mechanical commutation devices—the commutator. For eliminating this deficiency, electronic commutation (Stromag and Bosch companies) is suitable. The use of synchronous frequency controlled drives is conditioned on the development of new hard magnetic materials [6]. In addition to the known Alnico alloys and hard ferrites, cobalt-based alloys characterized by high permanent induction (0.8–1 T) and high density are being developed, while the demagnetization curve is almost straight. An Italian company Polymotor is producing ring drives for integrated spindle units on a base of $SmCO_5$ alloy. In an effort to reduce the consumption of rare earths and hence the cost of permanent materials, materials that do not contain rare earth are being developed. Mn-Al-C alloys are well known, as are materials containing CO, Cr, and Fe.

At the present time, the majority of manufacturers of integrated drive spindle units use asynchronic frequency controlled drives due to their advantages (**Table 2**).

For securing the drive parameters, it is necessary to choose a suitable frequency shifter, which processes the frequency of the 50 Hz network with an output frequency of up to 3000 Hz. They are thyristors or transistors with sinusoidal output. The main advantage of static converters compared to rotary converters is in

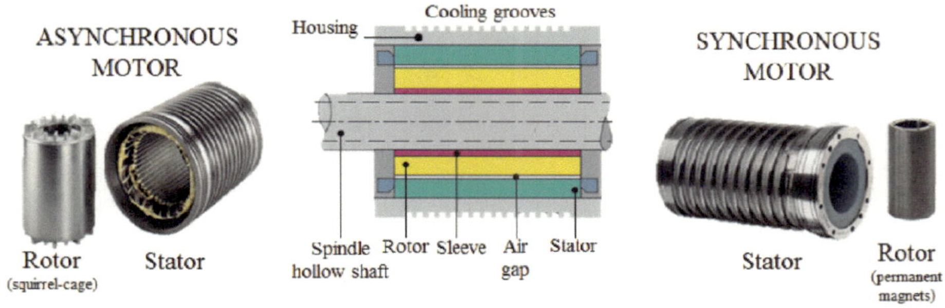

Figure 15.
Electric spindle motors (SIEMENS).

Lubrication method	Scheme	Describe	
Grease		Advantages	• Low price • General purpose application • Maintenance-free operation over long periods,
		Disadvantages	• Lower speed • It does not remove heat • Grease has smaller durability than oil
Oil-mist (oil-mist lubrication)		Advantages	• No worsening of lubricant quality • Water cannot get to the bearing area (oil mist forces it out)
		Disadvantages	• Ambient contamination • Oil quantity depends on temperature and viscosity
Oil-jet (oil-jet lubrication/ cooling)	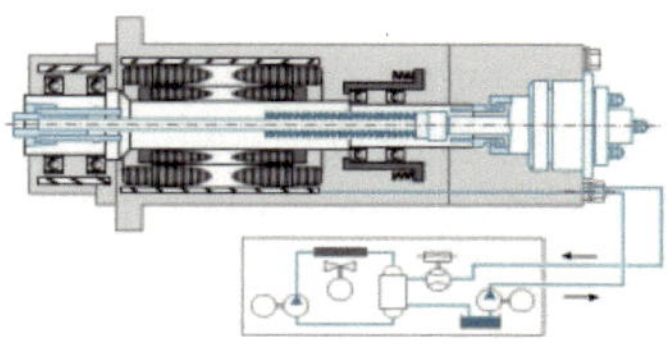	Advantages	• Stable bearing temperature • Water cannot get to the bearing area
		Disadvantages	• High friction moment • Higher price • Oil leakage at vertical application
Oil-air (oil-air lubrication)	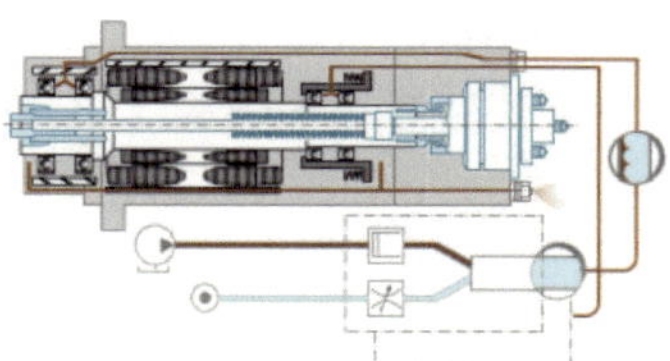	Advantages	• It is environmental friendly • Water cannot get to the bearing area • No worsening of lubricant quality • Stable bearing temperature • Low generation of heat from excess of lubricant
		Disadvantages	• High price • Difficult determination of oil quantity

Table 2.
Comparison of lubrication methods for spindle bearings [6].

continuous speed change control. Acceleration and braking work in a very short time without thermal load on the engine. There is no slip during braking, which is very advantageous for precise positioning of the spindle.

8. Peripheral

8.1 Clamping system

High-speed machining is associated with the development of new cutting materials such as cutting ceramics, synthetic polycrystalline diamond, and cubic boron nitride. In addition to the development of cutting materials with the new physico-mechanical and chemical properties, increased attention must also be paid to the optimization of machine geometry with regard to chip removal at high machined material volumes. It will be necessary to design new holder and clamper constructions in light of the frequency of revolution, rigidity, and the flow of cooling liquids.

In addition to the demands that are placed on clamping systems used in high-speed spindle units are the following constructional and technological requirements [15, 16]:

a. small clamp dimensions limited by spindle dimensions;

b. low weight, ensuring low centrifugal forces;

c. balancing, providing resistance to high frequencies; and

d. quick automatic tool or work piece exchange.

Interfaces are used for HSC machining centers: steep taper ISO, SK, BIG PLUS (taper 7:24) and especially short taper HSK (taper 1:10), Kennametal/Widia KMTS KM4X.

Clamping of the tool holder in the spindle cavity is usually done by pulling it in by means of restressed disc springs (**Figure 16**). The release is then a hydraulic cylinder (**Figure 17**). The advantage of the HSK type for high speed is that the centrifugal forces cause the collet to open, which rests on the internal cavity surface of the shank (**Figure 18**). Rotary turrets replace tools in less than 1 second and accuracy positioning is max. $\pm$ 3 μm.

Since HSC technology uses around 50,000 rpm, tools must have radial runout max. 0.003 mm and with interchangeable cutting plates (VRP) max. 0.01 mm. All tools used must be perfectly balanced.

In HSC technology, the following are most commonly used as tool holders:

- thermal fixture; and
- hydroplastic clamp.

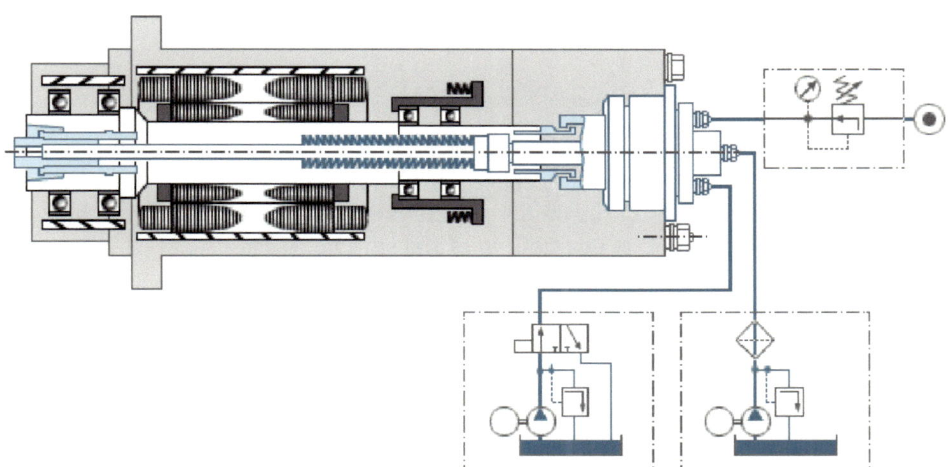

Figure 16.
Holder and clamper constructions (GMN).

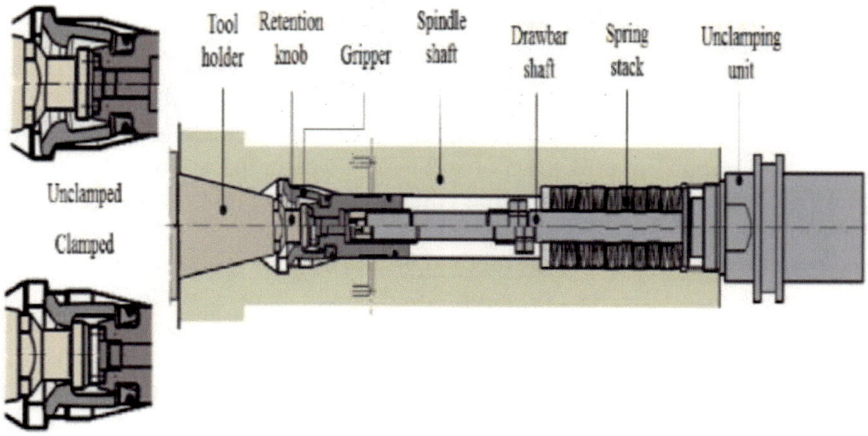

Figure 17.
End of spindle for clamping through the clamping shank [8].

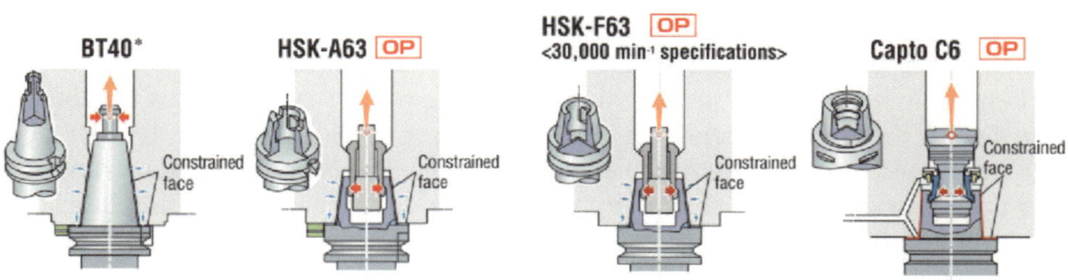

Figure 18.
Unsuitable and suitable clamping systems for high-speed spindles [DMG/Mori].

The thermal clamp allows quick clamping and unclamping of tools from the fast cutting steel, including sintered carbide. Tools are exchanged with a high-frequency generator that quickly warms up the tool holder and releases it tool. The following functions are used for the correct function of the tool: monitoring the tool holder contact in the spindle cavity, checking the temperature and force of the clamping system, checking the position of the clamping cylinder and the gripper, as well as checking the suction and temperature.

In the case of a hydroplastic clamp, the replacement is carried out using a hydraulic pump that squeezes the holder. Here, Pascal's law on the spread of uniform pressure is used, which ensures even clamping of the tool in the holder.

Advantages of these clamps:

- rapid shrinkage and release of the holder;
- the rigidity of the clamping ensures high quality of the machined surface;
- good bending and radial stiffness;
- clamping tools with shank h6;
- circumferential runout less than 3 μm; and
- use at maximum speed.

8.2 Lubrication and cooling system

In addition to the bearings themselves, the bearing parameters depend on the material and the quality of the surrounding parts, correct installation, and the choice of appropriate lubrication and cooling systems. These are lubrication and cooling of the contact point of the tool and work piece, lubrication and cooling of the bearings in the individual supports, and cooling of the motor and the headstock shell.

The correct choice of lubricant, method of lubrication, cooling liquid, method of cooling is as important for the proper operation of the bearing as the selection of the bearing and the design of the associated components. The methods used to lubricate and cooling the spindle bearings system at machine tools are shown in **Table 2**. Lubrication of bearings prolongs their life; it reduces the risk of their failures due to the mechanical damage at high speed; and it leads away generated heat. The lubrication method of spindle bearings at machine tools depends on the particular operation conditions.

The lubricating film thickness depends on the natural frequencies of rotation, operation temperature, and lubricant viscosity. In addition to the lubricant film thickness, it is necessary to assess the lubricant durability.

Grease for lubrication consists of 90% mineral oil or petroleum oil and 10% thickener. Lime soap, soda soap, lithium soap, or barium soap is used as the thickener. Grease durability depends on its quantity, sort, the bearing type, frequencies of rotation, and temperature in the assembled state. The bearings must be run in after their lubrication, and after a certain time period, they must be again lubricated. At running in, it is also necessary to take into account that grease can be well distributed on the whole bearing, which results in equalizing of temperatures generated by mechanic losses [6].

If the big accuracy is required at the spindle run, it is necessary to reduce heat. Passive friction moments that change to heat are influenced by the selected lubrication way and by the bearing design. The total passive friction moment is given:

$$M = M_0 + M_1 \quad [\text{N.mm}], \tag{20}$$

where M_0 is the friction moment dependent on the bearing design; and
M_1 is the friction moment dependent on loading (reaction).

The friction moment given by the bearing design and by the lubrication way is as follows:

$$M_0 = f_0 . 10^{-7} . (\nu . n)^{2/3} . d_{s^3} \ [\text{N.mm}], \tag{21}$$

where
f_0 is the coefficient given by the bearing design (0.7–12);
ν is the operation viscosity of oil or grease [$\text{mm}^2.\text{s}^{-1}$];
n is the frequency of spindle rotation [min^{-1}]; and
d_s is the mean spindle diameter [mm].

Lubrication by oil is used mainly in those cases where operation frequencies of rotation also require removal of generated heat from the bearing. At lubrication of the precise spindle bearings, it is necessary to use a small oil quantity to reach the high-quality bearing lubrication. The most widely used lubricating methods are:

- oil mist lubrication—the oil mist is produced in an atomizer and conveyed to the bearings by an air current. The air current also serves to cool the bearings and the slightly higher pressure prevents contamination from penetration;

Producer	Performance range [kW]	Revolutions range [min^{-1}]	Lubrication	Cooling	Drive	Technological operations
ENIMS	6.5	48–5200	Oil mist	Air liquid	AC	Tu
OMLAT	4.5–48.5	5000–40,000	Oil mist grease	Air	AC	Mi, Dr., Gr
IBAG	3–42	3000–80,000	Oil-air grease	Air liquid	AC	Mi, Dr
GMN	3–40	9000–60,000	Oil mist Oil-jet grease	Liquid	AC	Mi, Dr., Gr
FAG	2.5–20	20,000–45,000	Oil-jet minim. amount	Liquid	AC	Gr, Dr., Mi
SZM	10–15	20,000–75,000	Oil-jet grease	Air	AC	Mi, Gr
ITW	15	22,000–36,000	Oil mist	Air	AC	Mi, Gr
SKF	5.5–16	10,000–30,000	Oil mist Oil-air	Liquid	AC	Gr, Mi
MODIGS	1.4	70–2160	Grease	Liquid	DC	Gr, Mi
FORTUNA	0.45–15	12,000–18,000	Oil mist	Liquid	AC	Gr, Dr., Mi
SETKO	3.7	400–10,000	Grease	Liquid		Fr, Dr
PRECESI	0.17–6	7500–12,000	Oil mist	Liquid	AC	Fr, Dr., Gr

Machine units with AC integrated drive, DC, direct drive; Tu, turning; Mi, milling; Dr, drilling; Gr, grinding.

Table 3.
Components and peripheral devices used by selected manufacturers of electric spindles.

- oil-air lubrication—the oil is conveyed to the bearing in droplets by compressed air. The droplet size and the intervals between two droplets are controlled; and

- Oil-jet lubrication (cooling lubrication)—considerable amounts of oil are carried through the bearing by injection, the frictional heat generated in the bearing is dissipated. The cooling of the oil is achieved, for example, with an oil-to-air heat exchanger.

Table 2 describes various lubrication technologies and **Table 3** gives an overview of the individual components and peripheral devices used by selected manufacturers of electric spindles (**Table 3**).

9. Realized outputs

The spindle unit is determined by the structural parameters of the machine tool. In accordance with the growing requirements for production and precision of machine tools, the requirements for the design and technical execution of machine tool headstocks are increasing. The headstock of a machine tool is now a mechatronic, highly sophisticated system in which internal systems with external peripherals must interact. The design, research and development of new types of headstocks is not possible today without high-quality computing and simulation software, high-performance computing, testing equipment, and the necessary experience. For our headstock design, we have developed:

- special software that enables the calculation of the load, stiffness, and durability of rolling bearings used for the bearing of machine tool spindles [11–13]. Our methodology of calculation of bearing nodes associated from radial ball bearings angular contact is original. At the same time, the software enables to calculate the optimal distance of bearing supports with respect to overall maximum stiffness, running accuracy, and thermal expansion according to the arrangement of bearings in individual nodes as well as thermal expansion of the whole spindle bearing system; and

- special testing equipment for measuring the accuracy of running, temperature, and stiffness of bearing nodes made of radial ball bearings with angular contact. The experimental bench (**Figure 19**) enables the determination of required parameters of nodes arrangement by up to 5 bearings [6, 11]. Testing is carried out under different operating conditions.

The results of our work are designed more headstocks of machine tools. At our institute, we developed headstocks for CNC machine tools for the companies TOS Lipník and TOS Kuřim. The headstocks developed for SBL CNC lathes manufactured by the company Trens Trenčín deserve special attention [17]. The headstocks of the 300, 500, and 700 series developed at our workplace and the first lathes SBL was first time at the exhibition in Nitra 2000 and at the exhibition in Düseldorf 2004 presented. The SBL series lathes with the listed headstocks are still produced and are successful in the market.

Well known is our design of high-speed headstock with two drives. The advantage of the original solution is the possibility of using a high torque moment at low revs, as well as the principle of achieving high resultant revolutions of a double-mounted spindle driven by two drives [11]. The headstock is applied to a wood lathe in the company Šustrik.

An example of a new functional model of an electric headstock for a grinding machine is shown **Figure 20** [18]. It is a design of a headstock designed on the basis of modular components of an AC motor (stator, rotor, and metering system) and compact control system of the IMB Indramat drive. The functional model is preferably used in our laboratories for ultrasonic grinding.

When designing all headstocks, we use the V-2.16 headstock application software. The technical parameters of the rolling bearing nodes, such as axial and radial

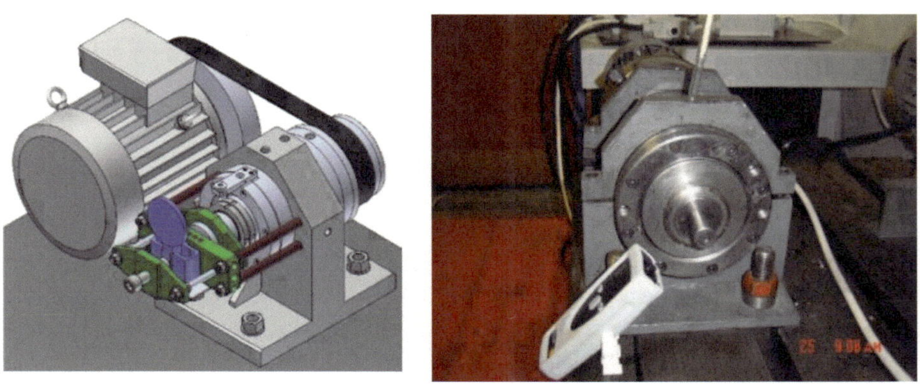

(a) (b)

Figure 19.
Laboratory for measuring: (a) testing equipment for measuring bearing nodes; and (b) test bench for testing functional models.

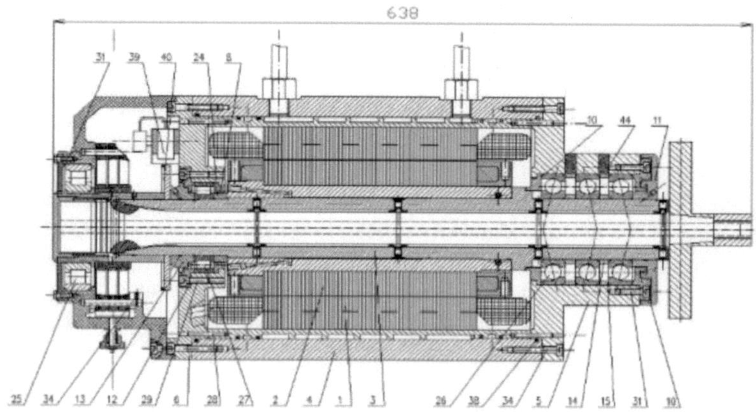

Figure 20.
Motor spindles for grinding machine tool [18].

stiffness, maximum speed, temperature, and running accuracy, are tested in our laboratories. Both the software and the experimental stand were developed at our workplace.

10. Conclusion

The technical level of fully automated flexible production systems has reached a degree at which accompanying working times and instruments are reduced to a minimum. Further increases in output are therefore possible by reducing the main production times.

This is possible by increasing cutting speeds—high-speed machining. Research in the field of high-speed cutting shows that, along with the reduction of lead times, cutting accuracy, productivity and machined surface quality are significantly improved.

In the chapter, requirements, characteristics, and development tendencies of the whole concept of construction of a machine, as well as construction nodes and elements of machine for high-speed cutting are described. With respect to the individual technological operations and the range and diversity of the required parameters, it is clean that at this time, it is not possible to design and universal machining unit—headstock. This requires a modular construction of the machining tool and individual peripheries that make possible a rapid change of the machining unit with the required revolution and performance characteristics.

The headstock is a determining structural node affected technological parameters machine tool. For high-speed machining, headstocks with built-in drive are used "Electrospindles."

The results of the analysis showed that electromagnetic and rolling spindles are used to accommodate the spindles of high-speed headstocks. Exceptionally with lower rigidity requirements, an aero-static bearing can also be used. The most widely rolling bearings used machine tool spindle support are nodes—formed from radial ball bearings with angular contact. They are reliable enough, cost-effective and, given the wide range of combinations, they can optimally meet the contradictory requirements of stiffness and maximum speed. Hybrid ball bearings are used for the highest rotational speeds but are very expensive.

In terms of drives, both single-direction and alternating drives can be used in the principle for integrated spindle units. Despite very good control properties, DC drives have known operational and technical drawbacks resulting from commutation devices. They are used less than AC drives. From the point of view of lubrication, the oil-air system is used for the highest rotational frequencies, while grease lubrication is still used for the lowest rotational requirements.

The headstock is a complicated mechatronic node with a system of internal elements and external peripherals. Designers must, in addition to complicated computer systems, perfectly master the demands placed on the headstock and the interoperability of individual elements and peripherals. New non-traditional materials (SI3N4, SmCO5 alloy) as well as progressive design technologies and design solutions are used to achieve the best technical parameters of these headstocks. At the end of the chapter, we present our results and experience in the design of headstocks of machine tools.

Acknowledgements

The research presented in this paper is an outcome of the project No. APVV-16-0476 "Research and development of the progressive design of the high speed rotor mounting in spinning machine" funded by the Slovak Research and Development Agency.

Author details

Ľubomír Šooš
Institute of Production Systems, Environmental Technology and Quality Management of the Faculty of Mechanical Engineering of STU in Bratislava, Bratislava, Slovakia

*Address all correspondence to: lubomir.soos@stuba.sk

References

[1] Schmockel D, Arnold W, Scherer J. Hochgeschwindigketsfräsen von Aluminium legierungen. VDI Zeitschrift. 1980;**122**(19):243-245

[2] Schulz H, Scherer J. Aktueller stnd des verbundforschurgsprojekts, Hochgeschwindigkeitsfräsen. Die Maschine. 1987;**10**:14-18

[3] Pasko R, Przybylski L, Slodski B. High speed machining (HSM)—The effective way of modern cutting. In: International Workshop CA Systems and Technologies. pp. 72-79. [Accessed: 24 April 2003]

[4] König W. Technologische Aspekte der Hochgeschwindigkeitzerspannung. Vorträg ansläβlich einer HGF—Tagung 17–18.4.1980. Eigenverlag der TH Aachen; 1980

[5] Available from: https://www.google.com/search?q=Skf+Motor+spindles

[6] Marek J et al. Designing of CNC Machine Tools. Praha: MM Publishing, s. r. o; 2010. p. 419 s. ISBN: 978-80_254–7980-3

[7] Lee D, Sin H, Sun N. Manufacturing of a graphite epoxy composite spindle for a machine tool. Annals CIRP. 1985; **34**:365-369

[8] Demeč P. Accuracy of Machine Tools and Its Mathematic Modelling. 1st ed. Košice, Vienala: Technical University in Košice; 2001. p. 146. ISBN: 80-7099-620-X (in Slovak)

[9] Available from: https://www.magnetic.waukbearing.com

[10] Available from: http://www.ibag.ch/en/downloads.html

[11] Šooš Ľ. Chapter: Radial ball bearings with angular contact in machine tools. In: Sehgal R, editor. Performance Evaluation of Bearings. Novi Sad, Croatia: Intech Prepress; 2012. pp. 49-92, 240. ISBN: 978-953-51-0786-6

[12] Šooš Ľ. New methodology calculations of radial stiffness nodal points spindle machine tool. In: International symposium on Advanced Engineering & Applied Management—40th Anniversary in Higher Education: Romania/Hunedoara/4–5 November 2010. Hunedoara: Faculty of Engineering Hunedoara; 2010. pp. III-99-III-104. ISBN: 978-973-0-09340-7

[13] Šooš Ľ. Approximate methodology calculations of stiffness nodal points. World Academy of Science, Engineering and Technology. 2011;7(80):1390-1395

[14] US Patent: US 6,422,757 B1 Active piezoelectric spindle bearing preload adjustment mechanism. July 23 2002

[15] Marek J et al. Designing of CNC Machine Tools. Praha: MM Publishing, s. r. o.; 2015. p. 727 s. ISBN: 978-80-260-8637-6

[16] Available from: http://www.gmn/en/downloads.html

[17] Šooš Ľ. Spindle–Housing system SBL 500 CNC. Eksploatacja i Niezawodnošč = Maintenance and reliability. 2008;**2**: 53-56

[18] Greguš Kolár J. High speed motor spindles for grinding (in Slovak). Bratislava: FME STU; 2015. p. 67

Section 2

Research and Development

Chapter 4

Analytical and Experimental Research of Machine Tool Accuracy

Peter Demeč and Tomáš Stejskal

Abstract

In this chapter, one of the ways of modelling the working accuracy of machine tools is elaborated. The method of constructing a numerical model based on transmission matrices is applicable in the field of virtual prototyping of modern machine tools. The results of numerical simulation of working precision are closely related to the machine stiffness in the machine workspace. A new way of measuring static stiffness can support numerical simulation results. Measurement of static stiffness is based on the measurement of the positioning accuracy of machine tools. The difference of the method lies in the load of the measured axis during the measurement. In this way, the stiffness of the movable machine partly relative to the base can be determined. The method for measuring static stiffness can advantageously be used in the development of new machine designs.

Keywords: machine tool, virtual prototyping, virtual machining, numerical experiment, static stiffness measurement

1. Introduction

Reducing the development time of modern machines is still a topical issue. Manufacturing experience is not sufficient to predict the characteristics of the machines being developed. Many solutions have to be tested and subsequently modified. This process requires experience but mainly time and cost. Computational technology in the form of virtual prototyping brings considerable acceleration of developing process. In addition to static parameters, the dynamic parameters of a virtual machine tool are now well verified. At the end of the development, a real machine tool is verified, which already requires considerably less modifications, as it was produced based on virtual testing of the virtual machine. Of course it is not possible to verify all the properties, but the system of virtual prototyping is constantly improving and expanding. This trend is confirmed by world trade fairs for machine tools (EMO). More and more sophisticated products supporting virtual prototyping are emerging on the market. Prototyping on the principle of virtual machining is a progressive direction. This type of analysis better simulates real production conditions and thus tests the designed machine using implemented mathematical models.

In this chapter, the practical application of virtual machining for virtual prototyping of machine tools will be explained, supplemented by the measurement of static stiffness of the machine tool. Typical sources of working inaccuracy are kinematic errors, thermo-mechanical errors, static and dynamic loads [1], a motion control method and control software [2]. In addition, the working inaccuracy is also determined by the appropriate choice of technological conditions and the required power performance.

The issue of dynamic software adaptation of Tool Center Point (TCP) is addressed [3]. In many works [4–8], the main influence on the TCP position has to do with the temperature factor. Stiffness experiments have shown that besides temperature, the clearance and the manifestation of contact stiffness between bearings and ball screw also play an important role in precision.

The stiffness of the machine tool has a significant influence on the accuracy of its work. Machine tools with a serial cinematic structure are working on the principle that the resulting relative movement of the tool and the workpiece is required to change the shape and dimensions of the workpiece and the results of the superposition of the individual machine nodes. Several mathematical and experimental methods are used to determine this stiffness. Consequently, many works deal with the calculation of stiffness by the finite element method [9, 10].

The classical static stiffness measurement is performed on a stationary machine most often at the tool clamping point relative to the workpiece clamping point. The load jig is gradually loaded and the working area subsequently uploaded in the direction of the machine axes. The deflection gauge measures the deformation at a given point in the load direction. Based on the measured values, the stiffness at the place location is determined. At the same time, deflections are also measured outside the load location. This determines the contribution of deflections to the stiffness change from individual machine components. Such measurement has great importance in prototype testing. Measurement of stiffness contributions reveals machine design imperfections. Some parts need to be redesigned to increase their stiffness. This achieves an acceptable machine stiffness. Normally stiffness measurements are rarely performed in in-process inspection.

The actual position of the moving part of the machine tool in relation to its base, even under static load, is influenced by three basic nonlinearities. These are friction, backlash and compliance [7].

The aim of this work is to design experimentally verified mathematical models that will significantly streamline, shorten and improve the quality of the work of the constituents. The aim of this chapter was to verify experimentally static stiffness using an interferometer, which opens new possibilities of evaluation. Interferometer measurement is used to evaluate positioning accuracy under the not loaded machine. The proposed stiffness measurement methodology with the use of an interferometer opens up new effects of the individual system pulses on the overall accuracy of the machine work. However, there are many nonlinear elements in this field, so the construction of a reliable mathematical model is considerably limited.

2. Theoretical fundamentals

The relative movement of the tool is determined by contributions from all parts of the kinematic chain. This kinematic chain is generally made up of serially connected machine components from the cutting tool to the workpiece holder. The relative movement between the tool and the workpiece is given by the precision of the kinematic components, the force deformations of the components and the random values that result from the nonlinear elements. The simplest solution for virtual motion is based on the kinematic motion of rigid components. The executive elements are characterised by nodes that can perform rotational or translational movement. In this way, the so-called ideal machined surface can be modelled. The surface is created as a trace of the ideal toolpath given by mathematical models.

To derive the mathematical model of the ideal tool trajectory, we will accept the following presumptions:

1. The relative movements of the machine's executive members will be examined as relative movements of the Cartesian coordinate systems linked with the workpiece, the individual executive members of the machine and its stationary nodes (e.g., the bed). All moving and stationary machine nodes involved in generating the resulting relative movement of the tool and workpiece will be called common how to model bodies.

2. The workpiece will have index 0, and its coordinate system S_0 (O_0, X_0, Y_0, Z_0) will be considered as stationary in the space.

3. For each of the model bodies, we assign the index i ∈ ⟨ 1; n⟩, where index 1 will have the model body immediately next to the workpiece and index n will have the model body that is the tool carrier. For other model bodies, we assign the increasing indexes in the direction from the workpiece to the tool.

4. The individual model bodies T_i will have the coordinate system S_i (O_i, X_i, Y_i, Z_i), whose axes are in the starting position (i.e., to the beginning of the machining), parallel to the corresponding coordinate axes of the workpiece $(X_i||X_0, Y_i||Y_0, Z_i||Z_0)$.

5. Model bodies can either stationary or can perform translational, respectively, rotational movement. One model body can perform only one of these movements and only in direction, respectively, around one coordinate axis of the previous model body.

6. The movements of all model bodies will follow the workpiece coordinate system.

Under these rules it is possible to derive relatively simple formulas for general application defining the position of any point $A \equiv [x_i, y_i, z_i]$ in the coordinate system $S_i(O_i, X_i, Y_i, Z_i)$, if you know the definition of his position $A \equiv [x_{i+1}, y_{i+1}, z_{i+1}]$ in the coordinate system $S_{i+1}(O_{i+1}, X_{i+1}, Y_{i+1}, Z_{i+1})$ and if the coordinate system given the previous coordinate system $S_i(O_i, X_i, Y_i, Z_i)$ performs any of the possible movements listed in the fifth rule.

In monograph [11] the fundamental mathematical relationships for the modelling of working accuracy of machine tools are derived with a serial kinematic structure of general shape. The mathematical model of the ideal tool trajectory is in the form of a matrix equation:

$$\{\mathbf{r}_0(t)\} = \left(\prod_{i=1}^{n}[\mathbf{R}_{i,i-1}(t)]\right)\{\mathbf{r}_n\} + \sum_{i=1}^{n-1}\left(\left(\prod_{j=1}^{i}[\mathbf{R}_{j,j-1}(t)]\right)(\{\mathbf{T}_{i+1,i}(t)\} + \{\mathbf{K}_{i+1,i}\})\right) + \{\mathbf{T}_{10}(t)\} + \{\mathbf{K}_{10}\} \tag{1}$$

where $\{\mathbf{r}_0(t)\}$ is the tool contact point position vector in the workpiece coordinate system (model body T_0),

$\{\mathbf{r}_n\}$ is the tool contact point position vector in the tool carrier coordinate system (model body T_n),

$[\mathbf{R}_{j,j-1}(t)]$ is the rotational motion transformation matrix of the model body T_i relative to the T_{i-1} model body,

$\{\mathbf{T}_{i+1,i}(t)\}$ is the linear motion transformation vector of the model body T_{i+1} relative to the T_i model body,

$\{\mathbf{K}_{i+1,i}\}$ is the starting position vector of the model body T_{i+1} coordinate system in the T_i model body coordinate system,

t is the time.

The final machining inaccuracy is determined by the sum of part deformation due to production forces and position inaccuracy of all model bodies (machine nodes) from tool to workpiece in a coordinate system of workpiece in time t that in mathematical language could be written in a form [12].

$$\{\boldsymbol{\Delta}(t)\} = \{\boldsymbol{\Delta}_0(t)\} + \sum_{i=1}^{n}\left(\left(\prod_{j=1}^{i}[\mathbf{R}_{j,j-1}(t)]\right)(\{\boldsymbol{\delta}_i(t)\} + [\boldsymbol{\varepsilon}_i(t)]\{\mathbf{r}_i(t)\})\right), \tag{2}$$

where the vector of deformations of the machined part is

$$\{\boldsymbol{\Delta}_0(t)\} = \{\boldsymbol{\delta}_0(t)\} + [\boldsymbol{\varepsilon}_0(t)]\{\mathbf{r}_0(t)\} \tag{3}$$

and the vectors

$$\{\boldsymbol{\Delta}_i(t)\} = \{\boldsymbol{\delta}_i(t)\} + [\boldsymbol{\varepsilon}_i(t)]\{\mathbf{r}_i(t)\} \tag{4}$$

represent the final position inaccuracies of active tool's point position caused by position inaccuracies of individual model bodies T_i expressed in the coordinate systems of these model bodies. The second part of Eq. (2) therefore represents a summary position inaccuracy of the active tool's point that is caused by position inaccuracies of all model bodies T_i and is reflected in the workpiece coordinate system.

The vector of linear inaccuracies of model body T_i in Eq. (2) defined relationship

$$\{\boldsymbol{\delta}_i(t)\} = \{\delta_{xi}(t)\ \delta_{yi}(t)\ \delta_{zi}(t)\}^T, \tag{5}$$

where $\delta_{xi}(t)$, $\delta_{yi}(t)$ and $\delta_{zi}(t)$ are linear inaccuracies in the direction of corresponding coordinate axes.

The matrix of angular inaccuracies of model body T_i is defined by the relationship

$$[\boldsymbol{\varepsilon}_i(t)] = \begin{bmatrix} 0 & -\psi_i(t) & \upsilon_i(t) \\ \psi_i(t) & 0 & -\phi_i(t) \\ -\upsilon_i(t) & \phi_i(t) & 0 \end{bmatrix}, \tag{6}$$

where $\varphi_i(t)$, $\upsilon_i(t)$ and $\psi_i(t)$ are angular inaccuracies (rotations about the axes X_i, Y_i, Z_i).

The position vector of active tool's point in coordinate system of model body T_i – vector $\boldsymbol{r}_i(t)$ in Eqs. (2) and (4) is defined by the relationship

$$\begin{aligned}\{\mathbf{r}_i(t)\} = &\left(\prod_{j=i}^{n-1}[\mathbf{R}_{j+1,j}(t)]\right)\{\mathbf{r}_n\} \\ &+ \sum_{j=i}^{n-2}\left(\left(\prod_{k=i}^{j}[\mathbf{R}_{k+1,k}(t)]\right)(\{\mathbf{T}_{j+2,j+1}(t)\} + \{\mathbf{K}_{j+2,j+1}\})\right) + \{\mathbf{T}_{i+1,i}(t)\} \\ &+ \{\mathbf{K}_{i+1,i}\}\end{aligned} \tag{7}$$

Virtual machining with thinking of acting forces calculating the inaccuracies of the position of all model bodies in the corresponding $t \in \langle O; T\rangle$ is mathematically defined by Eqs. (2)–(6). The result is a mathematical model of the real machined surface in the form of vector function

$$\mathbf{f}(t) = \mathbf{r}_0(t) + \boldsymbol{\Delta}(t). \quad (8)$$

3. Practical example

Practical implementation of the virtual machining on the virtual machine tool model is represented by the example of a virtual model of a machining centre with a horizontal spindle axis for machining non-rotating components, the design of which stems from the Box-in-Box concept and is complemented by a rotary table with the relvant construction dimensions (**Table 1**). This can be considered in a simplified

Dimension	a	b	B_4	B_5	h	h_3	h_4	h_5
Value (mm)	120	235	300	300	72	20	20	62
Dimension	H_3	H_4	H_5	L_2	L_3	L_4	L_5	p
Value (mm)	134	200	700	300	750	760	380	66

Table 1.
Design dimensions of the horizontal machining centre.

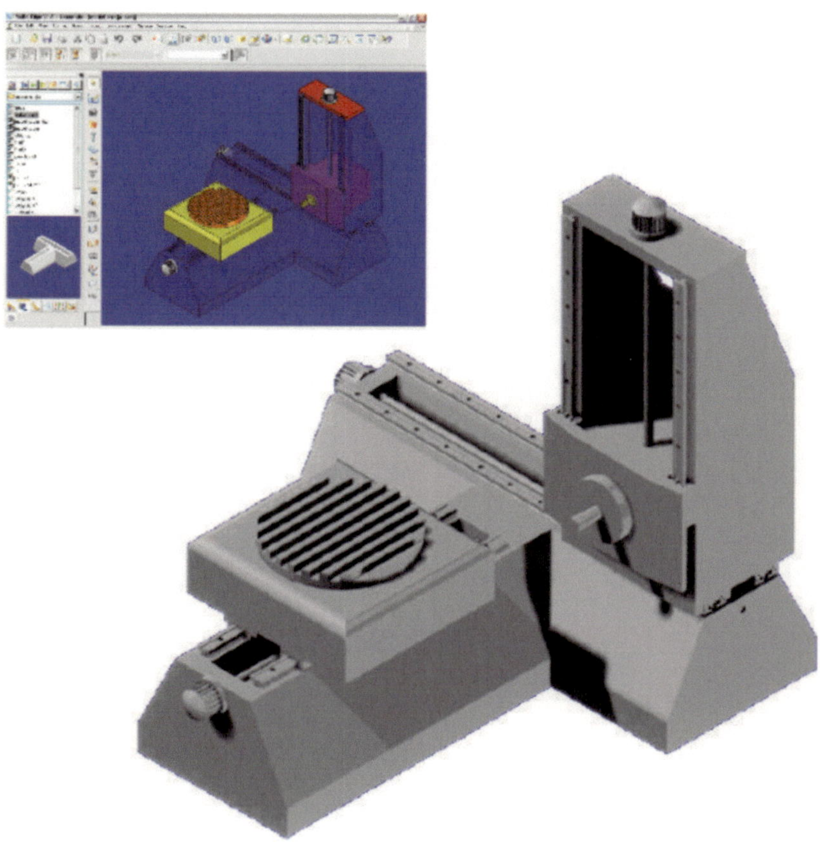

Figure 1.
Virtual model of the machining Centre.

version as a simple "indexing" table (allowing positioning of the workpiece for example by 90°) or as a variant with a controlled axis B (allowing machining when the workpiece is rotated about the vertical axis continuously).

The proposed machine (see **Figure 1**) has a work area defined by the maximum workpiece dimensions L_{omax} = 600 mm, B_{omax} = 500 mm a H_{omax} = 600 mm. The simplified machine model is illustrated in **Figure 2** and consists of the following

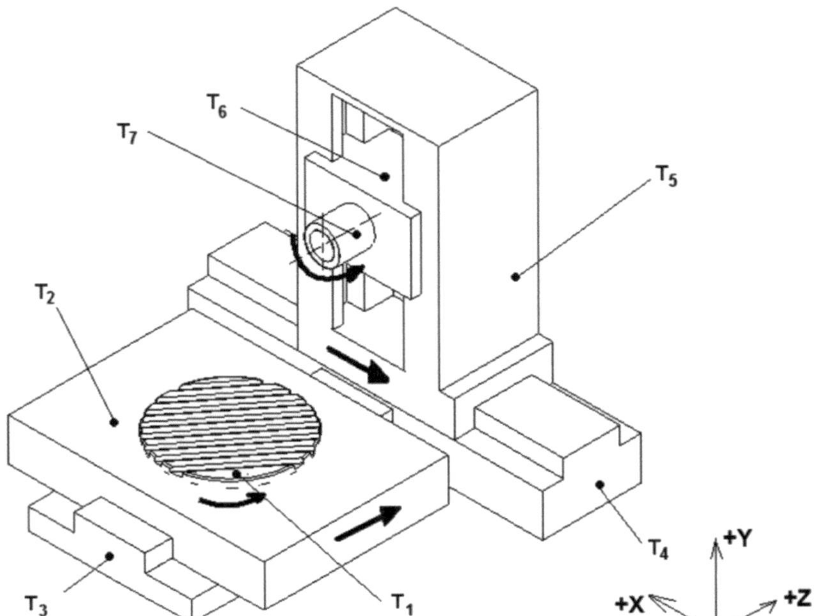

Figure 2.
Simplified computational model of the machining Centre with rotary table.

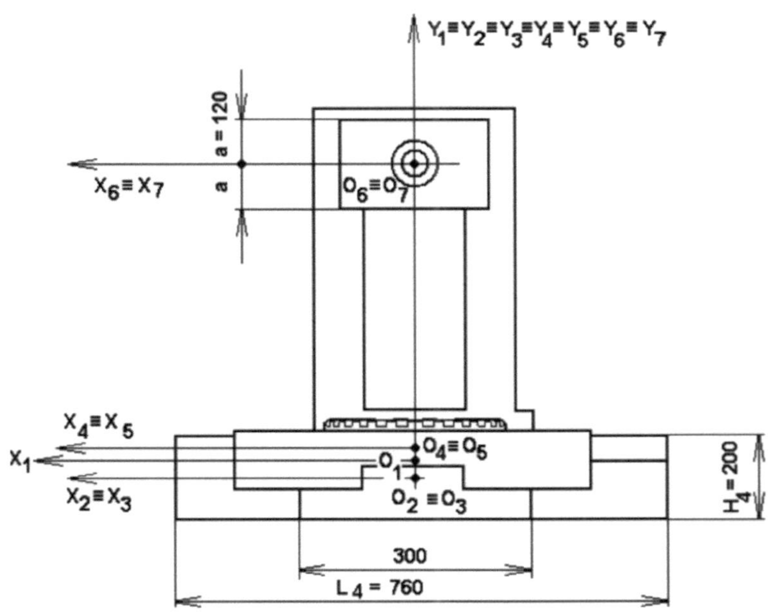

Figure 3.
Detailed computational model of the machining centre with rotary table (front view).

model bodies (gradually from the workpiece towards the tool): T_1, rotary table; T_2, table; T_3, bed (longitudinal part); T_4, bed (transverse part); T_5, stand; T_6, headstock; and T_7, spindle. The illustration also shows the basic layout of the machine's coordinate system. A detailed dimensional computational model of the machining centre is shown in **Figures 3–5**.

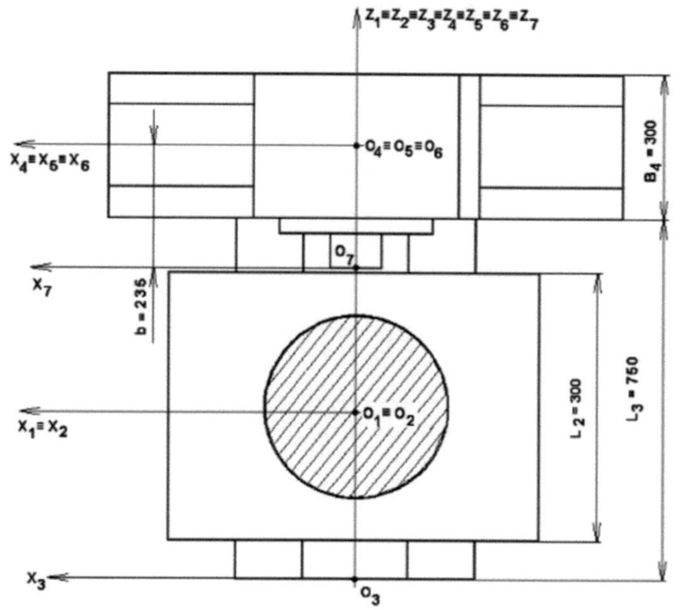

Figure 4.
Detailed computational model of the machining centre with rotary table (ground plan).

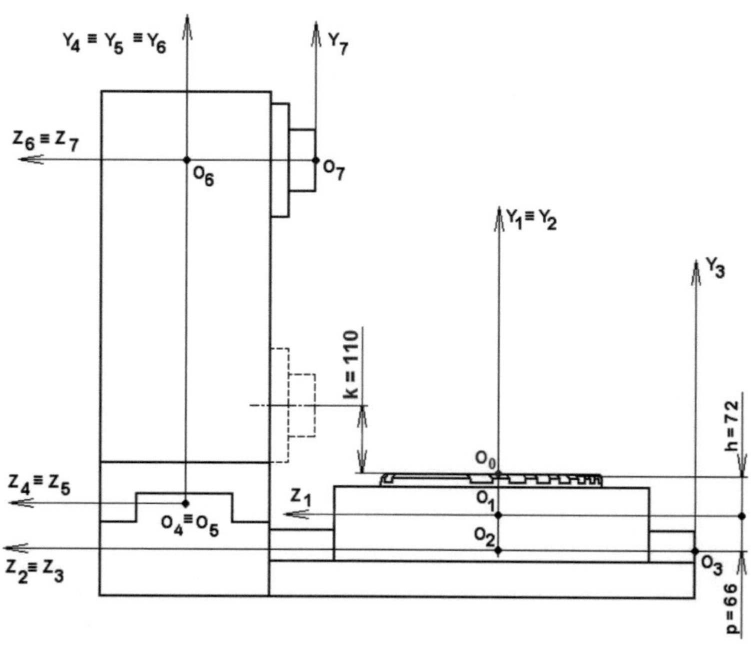

Figure 5.
Detailed computational model of the machining centre with rotary table (side view).

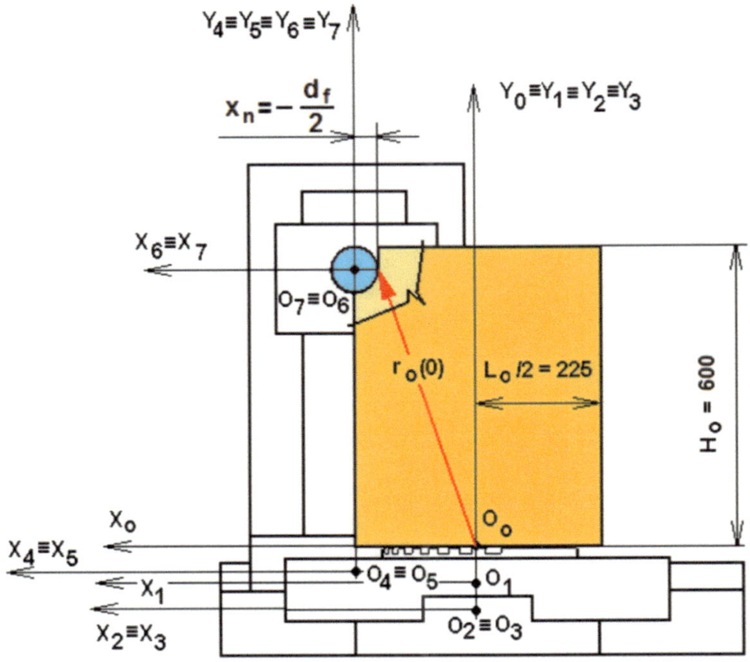

Figure 6.
Detailed computational model of the situation at the beginning of virtual machining (front view).

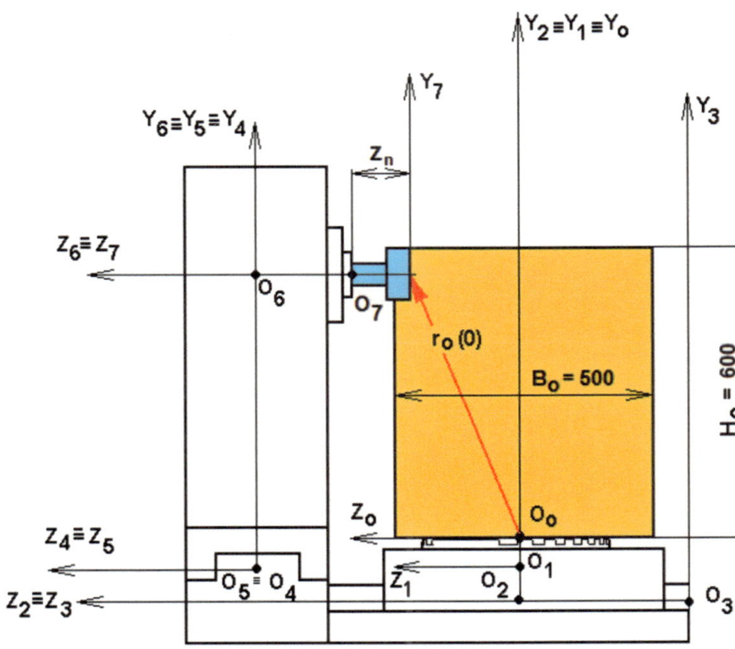

Figure 7.
Detailed computational model of the situation at the beginning of virtual machining (side view).

Based on the detailed computational models of each model body of the virtual machine, the corresponding transformation matrices and vectors were defined as follows:

The spindle: $[\mathbf{R}_{76}(t)]$, $\{\mathbf{T}_{76}(t)\}$, $\{\mathbf{K}_{76}\}$.
The headstock: $[\mathbf{R}_{65}(t)]$, $\{\mathbf{T}_{65}(t)\}$, $\{\mathbf{K}_{65}\}$.

The stand: $[\mathbf{R}_{54}(t)]$, $\{\mathbf{T}_{54}(t)\}$, $\{\mathbf{K}_{54}\}$.
The bed—transverse part: $[\mathbf{R}_{43}(t)]$, $\{\mathbf{T}_{43}(t)\}$, $\{\mathbf{K}_{43}\}$.
The bed—longitudinal part: $[\mathbf{R}_{32}(t)]$, $\{\mathbf{T}_{32}(t)\}$, $\{\mathbf{K}_{32}\}$.
The longitudinal table: $[\mathbf{R}_{21}(t)]$, $\{\mathbf{T}_{21}(t)\}$, $\{\mathbf{K}_{21}\}$.
The rotary table: $[\mathbf{R}_{10}(t)]$, $\{\mathbf{T}_{10}(t)\}$, $\{\mathbf{K}_{10}\}$.

The virtual workpiece was designed as a quad shape body with dimensions L_o = 450 mm, $B_o = B_{omax}$ = 500 mm and $H_o = H_{omax}$ = 600 mm. The position of the workpiece on the table is symmetrical, and the vertical planes of symmetry of the workpiece and the table are identified.

In virtual machining, the simultaneous machining of two vertical planar faces of the workpiece by the front and cylindrical peripheral surfaces of the tool—the front cylindrical cutter—was considered. It is considered an unsymmetrical down-milling. The tool will be the front cylindrical milling cutter R215.59–06, the diameter d_f = 20 mm and the number of teeth z_f = 4. Milling depth h_f = 10 mm, milling width b_f is equal to half the diameter of the cutter d_f, machining with motion of the headstock (direction Y) is from the top down, feed rate f_Y = 0.09 mm (feed rate on the milling tooth), cutting speed v_c = 170 m·min^{-1}, and spindle speed is adjusted to n_v = 2700 min^{-1}. When looking from the machine stand to the workpiece, we cut the front of the workpiece over the entire height machining starting from the top right corner of the workpiece (see **Figures 6** and 7).

4. Virtual machining result

The machining of the respective planar faces of the workpiece by the front cylindrical milling cutter begins at the coordinate $y_O(0)$ = 590 mm and ends at $y_O(T) = k$ = 110 mm, resulting from the construction of the headstock and spindle. Thus, the headstock travel at time T is 480 mm, with the total machining process lasting at selected feed rates, and the number of milling teeth and spindle speed $T = 480/16.2 \approx 29.63$ s. If we divide the total machining time into, for example, 10 equal sections, Δt = 2.963 s; the headstock path is simultaneously divided into 10 equal sections of 48 mm in length. In these positions of the headstock, the corresponding numerical simulations are performed.

The power ratios for machining (see **Figure 8**) were simulated on the basis of the structural equation for the tangential component of the cutting force (cutting resistance) according to [13]

$$F_c = 682 h_f^{0.86} b_f z_f f_Y^{0.72} d_f^{-0.86} \ (\text{N}) \tag{9}$$

where the appropriate dimensions are set in millimetres (mm). On the basis of Eq. (9), according to [13], the components of the cutting force (cutting resistance) in the directions of the individual coordinate axes are valid for asymmetrical down-milling:

$$\left.\begin{aligned} F_X &= (0.60 \div 0.90) F_c \\ F_Y &= (0.45 \div 0.70) F_c \\ F_Z &= (0.50 \div 0.55) F_c \end{aligned}\right\}, \tag{10}$$

Due to the proportions of the design of the individual model bodies of the machine, only the deformations of the stand were considered in the other calculations. For numerical simulations, relationships (10) were used for the most unfavourable values; therefore, the respective components of the cutting forces (resistances) are for down-milling:

$$\left.\begin{aligned} F_X &= 0.90\,F_c = 3386.0563\ \text{N} \approx 3386\ \text{N} \\ F_Y &= 0.70\,F_c = 2633.5993\ \text{N} \approx 2634\ \text{N} \\ F_Z &= 0.55\,F_c = 2069.2566\ \text{N} \approx 2070\ \text{N} \end{aligned}\right\}, \tag{11}$$

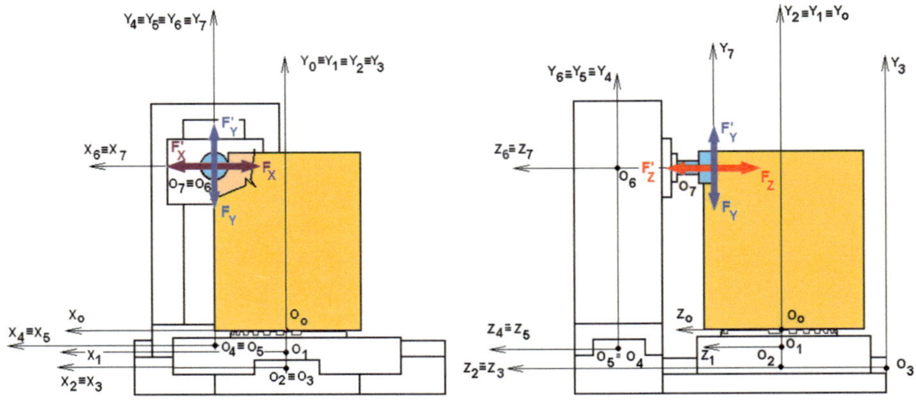

Figure 8.
Layout of cutting forces and resistances for virtual machining.

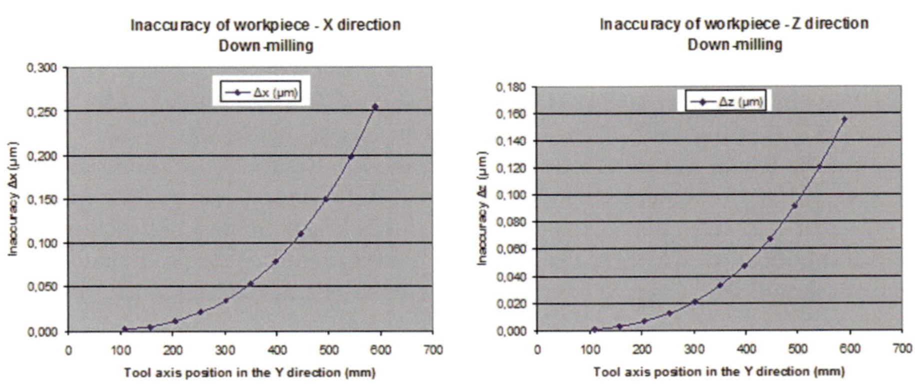

Figure 9.
Graphical representation of workpiece inaccuracies.

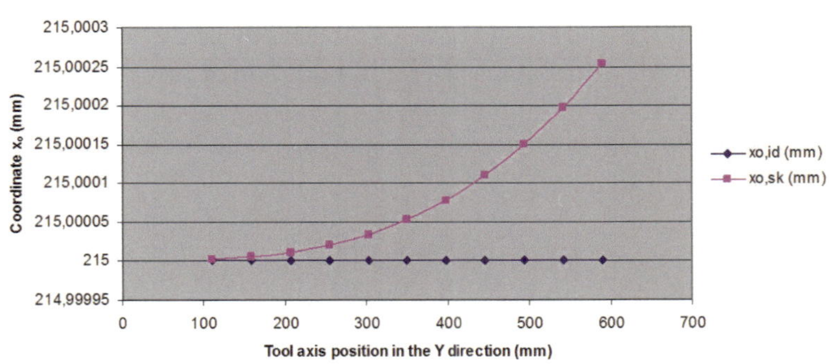

Figure 10.
The real coordinates of the machined surface in X direction by down-milling.

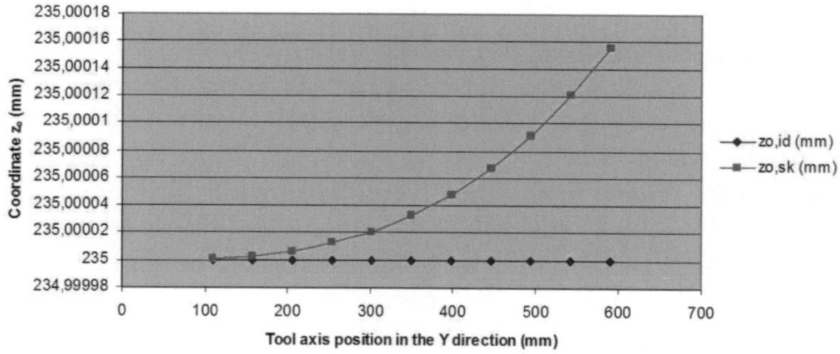

Figure 11.
The real coordinates of the machined surface in Z direction by down-milling.

Position of the headstock i	Down-milling							
	t	$y_{Oi}(t)$	$\Delta x_{Oi}(t)$	$x_{Oi},i_d(t)$	$x_{Oi},_{sk}(t)$	$\Delta z_{Oi}(t)$	$z_{Oi},_{id}(t)$	$z_{Oi},_{sk}(t)$
	(s)	(mm)	(μm)	(mm)	(mm)	(μm)	(mm)	(mm)
1	0	590	0.254	215	215.00025	0.155	235	235.00016
2	2963	542	0.197	215	215.00020	0.120	235	235.00012
3	5926	494	0.149	215	215.00015	0.091	235	235.00009
4	8889	446	0.110	215	215.00011	0.067	235	235.00007
5	11.852	398	0.078	215	215.00008	0.048	235	235.00005
6	14.815	350	0.053	215	215.00005	0.032	235	235.00003
7	17.778	302	0.034	215	215.00003	0.021	235	235.00002
8	20.741	254	0.020	215	215.00002	0.012	235	235.00001
9	23.704	206	0.011	215	215.00001	0.007	235	235.00001
10	26.667	158	0.005	215	215.00000	0.003	235	235.00000
11	29.63	110	0.002	215	215.00000	0.001	235	235.00000

Table 2.
Virtual machining results with down-milling.

Some results of numerical experiments are shown in **Figures 9–11**; numerical values are given in **Table 2**.

5. Experimental measurement of milling machine stiffness during X-axis positioning

Inspiration to modify the commonly used extended static stiffness measurement method resulted from the significantly different experimentally measured static stiffness values of the new loading method compared to the standard stiffness measurement method. Static stiffness measurements in this experiment were performed under the axis load immediately after reaching the desired position. From the known force and deflection, it is possible to determine the stiffness of the table relative to the base at a given programmed linear unit position. Thus, the stiffness measurement results more faithfully reflect the actual ratios that occur during machining. This note is important in that relatively slow changes, such as a change in the thickness of the oil layer in the contact surfaces, may occur after the movement has stopped, which may affect the static stiffness.

An important observation from the measurements is that the measured static stiffness of the table greatly depended on the previous operation of the table and the way it was loaded. The classical stiffness measurement showed up to three times higher stiffness values than the modified method using identical tools and conditions. The reasons for such a considerable difference in stiffness are given by the structural arrangement and the effect of nonlinearities on the contact surfaces of the cross-table components. During the measurement, all machine parts were stationary. This modified view of the static stiffness can be used to change the design philosophy of the machine tool. This is applicable in engineering practice, especially in the field of machine tool design, which will ensure higher machining accuracy under comparable conditions. In the experiments performed, the Renishaw XL80 laser interferometer was used to measure deformation and displacement. Measuring accuracy of the manufacturer guarantees better than 0.5 μm/m.

5.1 Arrangement of experiments

For experimental static stiffness measurement, the Kondia B 640 CNC Vertical Milling Machine was used as a production machine with three fully controlled axes [14].

5.1.1 First experiment: gradual loading of table positions

During gradual loading of table positions, the work procedure consisted of alternating cycles with and without the load. The reason for such a procedure was to estimate the thermal expansion of the table against the base during one cycle. One cycle consisted of the table's gradual positioning of nine positions in the X axis (**Figure 12**). Using the laser interferometer, an exact position was measured.

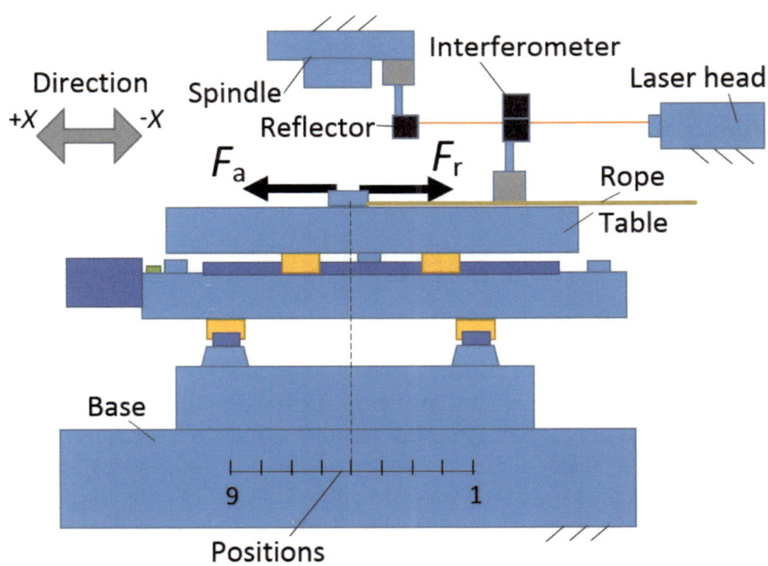

Figure 12.
Measurement sequencing in direction of the X axis on the cross-milling table.

From this, stiffness was determined in each position in both directions of the table starting with the given position (**Figure 13**). At each position, the average offset from the slope can be calculated.

As shown, the stiffness measured by this method is approximately the same in either direction. The highest stiffness was detected around position 8. This fact is consistent with the machine design since the lowest matrix and motor distance is at this position (**Figure 12**). The screw in this position is least involved in the stiffness deterioration.

5.1.2 Second experiment: static loading of table positions

The static loading of the table was performed to verify the continuous loading method. The working procedure consisted of moving the table to the measured position. In the given position, it was gradually loaded and relieved by the F_r force. In each position, the loading and unloading cycle was repeated twice. The reason was that only at the second cycle the backlash from the load was determined in the given direction. The stiffness of the system has been determined from the charted slope (**Figure 14**). This method of stiffness measurement can practically be considered a classic one.

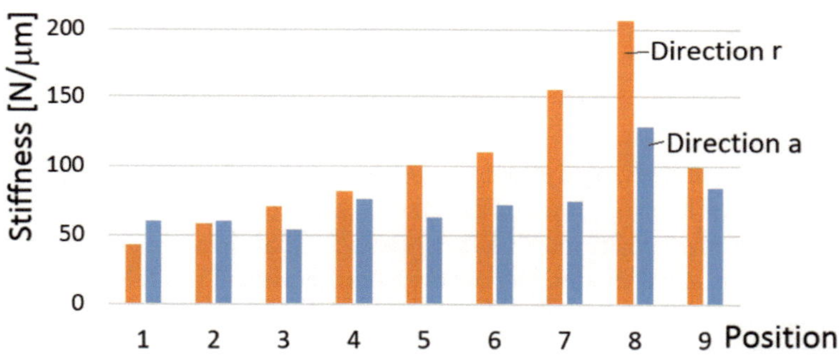

Figure 13.
Measured stiffness of the shift in the a and r directions, respectively.

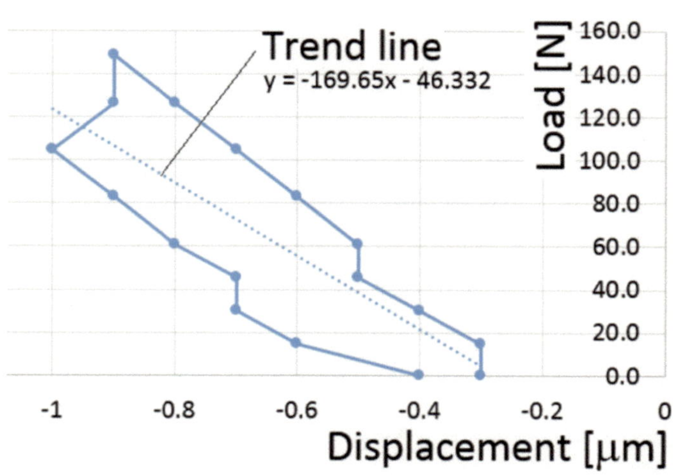

Figure 14.
Stiffness measurement in position 8 (second loading). The slope shows stiffness (169 N/µm).

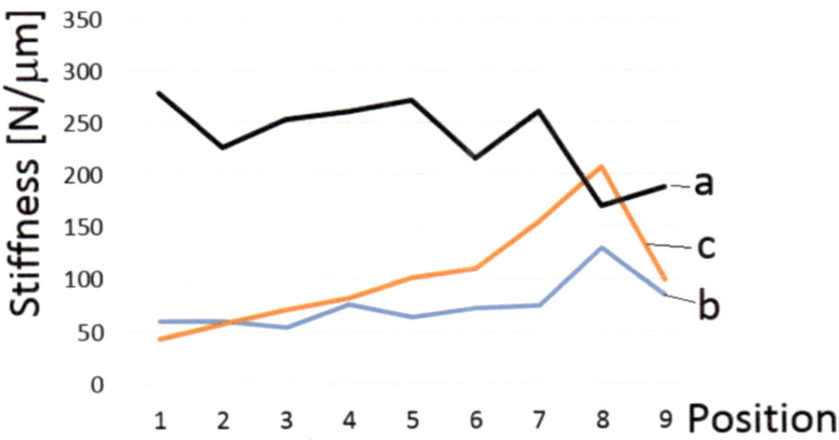

Figure 15.
Size comparison of (a) the static stiffness, (b) the shift stiffness in direction a *and (c) the shift stiffness in direction* r.

The course of static stiffness in individual positions is shown in **Figure 15**. The resulting course was compared with the results of the first experiment. As the graph clearly shows, the static stiffness is considerably different from the shift stiffness.

6. Discussion

As the experiments show, the actual position of the table depends on the previous method of loading. Immediate static load can only be given a partial credit for the assumed position. This fact places special demands on how to compose a mathematical model. In common mathematical models, the input values translate into a clear result. This is not the case. For that reason, we did not proceed with creating the mathematical model. Only a schematic model has been created (**Figure 16**) [14].

The basis of the model is the source of resistance to movement. These are friction, deformation of the contact surfaces and elastic deformation of the machine

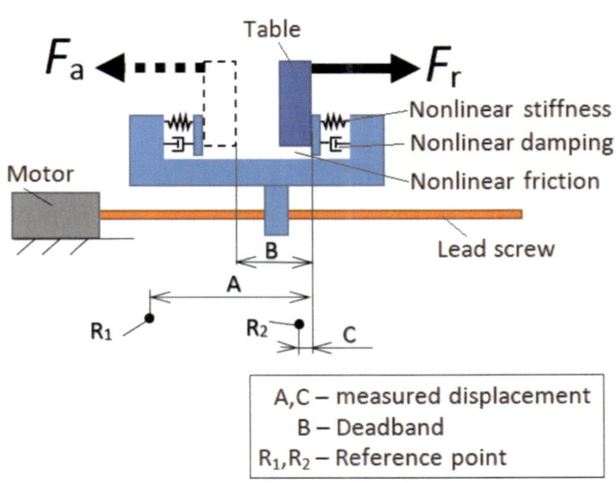

Figure 16.
Machine positioning model.

structure. All three sources cause nonlinearity between the load and the position change.

As follows from the experiments, the classic machine stiffness measurement does not detect design deficiencies in terms of drive quality and the machine's alignment. The measured stiffness values are in the order of three times the values measured when the machine is shifted to the measured position. Yet, the method of shifting the movable parts under load clearly reflects the real state during machining better.

The stiffness analysis shows that if the load direction is not changed, the stiffness will be higher. This is related to the definition of the backlash and the direction of the previous load in the individual parts of the machine.

The purpose of the construction of the mathematical model is based on the idea of improving the design of the machine in terms of work accuracy of production. There are two possible procedures. The first is to minimise the adverse effect of nonlinearities in the construction nodes. This can be achieved, for example, by selecting preloaded connections, reducing the number of kinematic elements to a minimum and the like. The second procedure is control of nonlinear states. This procedure requires feedback that is provided by sensors, for example, temperature, vibration sensors and force and deflection sensors. The last two sensors can be used to the stiffness measure of the measured point at a given time and space.

Theoretically, there is no analytical method for calculating nonlinearities. For the first time, French mathematician Henri Poincaré demonstrated it in the task of solving the movement of the three heavenly bodies (three body problem) at the end of the nineteenth century. However, the awareness of this fact was very gradual. It was not until the development of computer technology and chaos theory in the 1970s of the last century that nonlinear problems began to be solved by numerical simulation. Such a nonlinear model gives a solution, but the trajectory of the output functions is uncertain. It moves with some probability that the result is in some interval. The system works very much like the weather forecast.

7. Conclusion

The stiffness measurement by laser interferometer opens up new possibilities of evaluation, not exactly feasible under classic measurement with dial micrometre indicators. The results obtained by measurements point to the fact that the static stiffness depends both on the previous method of loading and the direction of the start of the measured position. The proposed shift stiffness measurement methodology allows for a better assessment of the machine's working ability. It also allows for the detection of faults caused by, for example, incorrect assembly of machine components. The method also points to structural design deficiencies that would not be detectable from a classic stiffness measurement.

The comparison of the theoretical inaccuracy resulting from the design of the machine components to some extent is also related to the experimental results of the measurement of the static stiffness at different axis positions. Experiments show that the stiffness change is also related to the location of the nut and bolt (**Figure 13**, position 8). This is the influence of design accuracy, which was theoretically described in the introductory chapters. In addition, the previous working action and the amount of load, which is also a design matter, also affect the stiffness and hence the accuracy. However, there are many nonlinear elements in this field, so the construction of a reliable mathematical model is considerably limited. When designing the design of good machines, it is advisable to keep these effects in mind.

Acknowledgements

This work was supported by the Slovak Research and Development Agency under the Contract no. APVV-18-0413: Modular architecture of structural elements of production machinery.

Author details

Peter Demeč and Tomáš Stejskal*
Department of Manufacturing Machinery, Faculty of Mechanical Engineering, Technical University of Košice, Košice, Slovakia

*Address all correspondence to: tomas.stejskal@tuke.sk

References

[1] Chen XB, Geddam A, Yuan ZJ. Accuracy improvement of three-axis CNC machining centers by quasi-static error compensation. Journal of Manufacturing Systems. 1997;**16**(5): 323-336

[2] Lin M-T, Shih-Kai W. Modeling and improvement of dynamic contour errors for five-axis machine tools under synchronous measuring paths. International Journal of Machine Tools and Manufacture. 2013;**72**:58-72

[3] Mayr J et al. Comparing different cooling concepts for ball screw systems. Proceedings of ASPE Annual Meeting; 2010

[4] Mayr J et al. Thermal issues in machine tools. CIRP Annals - Manufacturing Technology. 2012;**61.2**: 771-791

[5] Brecher Chr, Wissmann A. Modelling of thermal behaviour of a milling machine due to spindle load. 12th CIRP Conference on Modelling of Machining Operations. Vol. 2; 2009

[6] Wang Y. et al. compensation for the thermal error of a multi-axis machining center. Journal of Materials Processing Technology. 1998;**75**(1):45-53

[7] Donmez MA, Hahn M-H, Soons JA. A novel cooling system to reduce thermally-induced errors of machine tools. CIRP Annals - Manufacturing Technology. 2007; **56**(1):521-524

[8] Gebhardt M et al. High precision grey-box model for compensation of thermal errors on five-axis machines. CIRP Annals - Manufacturing Technology. 2014;**63.1**:509-512

[9] Huang DT-Y, Lee J-J. On obtaining machine tool stiffness by CAE techniques. International Journal of Machine Tools and Manufacture. 2001; **41.8**:1149-1163

[10] Altintas Y et al. Virtual machine tool. CIRP Annals - Manufacturing Technology. 2005;**54**(2):115-138

[11] Demeč P, Svetlík J. Virtual Prototyping of Machine Tools. 1st ed. RAM-Verlag: Lüdenscheid; 2017. p. 158. ISBN 978-3-942303-61-3

[12] Demeč P, Svetlík J. Virtual machining and its experimental verification. Acta Mechanica Slovaca. 2009;**13**(4):68-73

[13] Lipták O et al. Production Technology - Machining. 1st ed. Bratislava: ALFA; 1979. (in Slovak)

[14] Stejskal T, Svetlík J, Dovica M, Demeč P, Kráľ J. Measurement of static stiffness after motion on a three-axis CNC milling table. Applied Sciences. 2018;**8**(1):15. DOI: 10.3390/app8010015

Chapter 5

Geometric Accuracy, Volumetric Accuracy and Compensation of CNC Machine Tools

Jiri Marek, Michal Holub, Tomas Marek and Petr Blecha

Abstract

The production of geometrically and dimensionally defined workpieces is what the user expects from a machine tool. Deviations from these prescribed dimensions and geometry are due to machine inaccuracies. Therefore, it was necessary to develop tests and tests on the properties and parameters of machine tools that can detect these. Every new machine tool undergoes these tests.How to perform and evaluate these tests is determined and recommended primarily by standards and regulations. When testing the properties of machines, it is not only about knowing and knowing how to measure machines, but also how I can analyze and apply the obtained results. Is it necessary to do a mechanical intervention of the machine or is it enough to compensate the software?

Keywords: geometric accuracy, volumetric accuracy, compensation, machine tools

1. Introduction

The production of geometrically and dimensionally defined workpieces is what the user expects from a machine tool. Deviations from these prescribed dimensions and geometry are due to machine inaccuracies. Therefore, it was necessary to develop trials and tests of machine tool properties and parameters that can detect these errors. Every new machine tool, a newly developed machine, or a machine overhauled is subjected to these tests [1].

Testing of machine tools is an important part of the product life cycle-machine tool. Tests of machine tools can be divided into three groups. The first group of tests is associated with a contractual obligation between the seller and the buyer of the machine. They are, therefore, a part of the contract. Acceptance tests usually take place in two steps—first, directly at the machine manufacturer and then, after the machine is assembled, at the customer. These tests aim to verify the declared properties of the machine. The prototype tests serve to verify the properties of newly designed and manufactured machines. Prototype tests extend the acceptance tests with a series of measurements to provide important information, especially to machine designers. The proposed and expected properties of the new product are examined and the unknown properties, which cannot be expected when the product is being developed, are revealed. Statistical acceptance (process competence test) is used for exacting customers, where it is necessary to maintain the quality of the workpiece in the long term [2].

How to perform and evaluate these tests is determined and recommended primarily by standards and regulations. When testing the properties of machines, it is not only about knowing and being capable of how to measure machines (what kind of equipment to use, what method and procedure), but also how to analyze and apply the results in future. Is it necessary to do a mechanical intervention into the machine or is it sufficient to compensate the machine software? [1].

The inspector should be able to answer these and other questions related to machine tool diagnostics. Machine diagnostics is not only a knowledge of the measurement method, but also a set of knowledge that the inspector must know. The first is the knowledge of the measuring equipment itself and its management, monitoring its properties, accuracy, and ensuring a regular calibration (if necessary). Next, it is the knowledge of working with these devices (procedures) and what standards and regulations apply to the measured quantity, the machine, and the device itself. However, it is also important to know the measured machine, without which we cannot adequately perform diagnostics and propose suitable measures to improve the accuracy of the machine [1].

The publication [3] describes the effects of an improperly selected method of measuring the volumetric accuracy of a machine tool. Various methods of placing the temperature sensors on the machine were carried out. These are then reflected in the size of individual machine errors, but also in the resulting volumetric accuracy in the range of 8–12%. This is an example of a different approach to measuring of volumetric accuracy, which is, in this case, affected by the human factor.

2. Effects influencing CNC machine tool operation

The machine tool must be seen as a technical system, which must always be considered in a comprehensive way, with all the impacting effects. In operation, the CNC machine tool is influenced by a number of effects. By this, we understand the effect not only of the ambient where it is installed, but also the influence of the operator on the machine itself and its impacts on the ambient. These influences affect the properties that all machine tool users call for, namely run stability, repeated machining accuracy, and trouble-free operation. We must assess machine tools in a comprehensive, hierarchical, and structured way. The deviations in the dimensions of the machined component provide the user with direct information on the accuracy of the parts from which the machine is assembled, on the care devoted to the assembly and, last but not least, on its construction. The workshop environment where the machine is installed affects the machine tool by [4]:

- vibrations;
- impurities;
- heat.

On the other hand, the machine can have the same effects on the environment. The machine can cause vibrations (not common), exhaust gases from the supply of coolant and cutting fluid to the cutting site and can also cause ambient warming. By impurities we do not mean coarse dirt and excessive dust, but the standard ambient of normal workshop operation. Heat flow and radiation from the ambient have an immediate effect on the machine installation site and can adversely affect the machine operation. Coldness or sudden temperature changes are equally unfavorable. In cases where this does not impede the operation of the machine (e.g., thermal protection failure, functionality of motion mechanisms) and the temperature changes (sudden temperature difference) are not too high, the machine can be operated satisfactorily. This state can be compared to a temperature steady state (tempered state). Therefore, manufacturers usually report the temperature range at which their machine operates. Rather, a sudden change in the temperature field is detrimental [4].

In addition to these external effects, several factors, referred to collectively as production accuracy (production uncertainty), affect the operation and, in particular, its machining accuracy. When machining a workpiece over time, its dimensions vary within or outside the given and permitted limits. Workpiece dimensional variations are caused by three main factors affecting the machine tool and the manufacturing process [4]:

- temperature influence;
- static rigidity of the machine-tool-workpiece system;
- dynamic compliance of machine-tool-workpiece system.

Every CNC machine tool is exposed to temperature effects, both even and uneven, during its operation and also in its sleep mode. Due to this temperature effect, temperature deformations arise which lead to a change in the position of the workpiece relative to the tool and thus to inaccuracies. This will be striking if we are focused on the stability of the machined dimension in case of a smaller series of workpieces, respecting the shape and position errors defined on the machined parts. The causes of heating up the individual parts of the machine tool can be found either in the machine itself (passive resistors in the motion axes or the cutting process itself) or outside it. The thermal stability of machine tools today is one of the most important factors for maintaining the specified tolerances on the workpiece [5].

Almost all the mechanical work that is done in the cutting process turns into heat. In addition, losses occur in the machine motion groups. Heat is dissipated from the place of origin (cutting process or in drives, guides) by [5]:

- conduction;
- convection;
- radiation.

Heat dissipates from the cutting process by:

- chip;
- workpiece;
- tool;
- ambient.

It follows that almost all the heat is stored in the machine tool and must be dissipated or stabilized. Uneven heating up of machine tool parts can occur, which can lead to thermal expansion and deformation. This results in fluctuations of workpiece dimensions and tolerance variations in shape and position. All temperature effects cause a temperature increase during machine tool operation, which then stabilizes at a certain value—the so-called steady temperature, which is different for each machine. Therefore, some manufacturers insist on this condition and then recommend machining. However, they must ensure that there is no sudden change in temperature. The harm caused to the machining process may not be the temperature itself, but rather harms of temperature changes during machining. For this reason, in addition to efficient cooling, some manufacturers also heat their machines [5].

This state is called a thermally stabilized machine tool. The cold machine tool heats up slowly, because we cannot achieve smooth operation and even workload of the machine tool at the beginning of machining. This is because machining must often be interrupted and this causes cooling. Therefore, at first, the machine is thermally stabilized by heating to the operating temperature and then by controlling and maintaining its temperature. Our aim is that, in spite of the thermally stabilized state of the machine, the changes in temperature and its manifestations of thermal deformation could affect as little as possible the position of the tool relative to the workpiece and thus the machining accuracy by [5]:

- selecting a thermo-symmetrical machine design;
- increasing the efficiency of all nodes and elements, thus minimizing losses that change into heat;
- placing heat sources efficiently so that they do not affect the design of the machine;
- dissipating the heat by cooling, chip removal, or by dimensioning the surfaces for efficient heat dissipation;
- compensating the machine;
- checking the air flow and its temperature, or shielding the external thermal radiation.

Undesirable and harmful side effects of time-varying loading can be vibrations, and thus also the accompanying phenomenon of these vibrations—noise of the machine or its parts. Vibrations deteriorate the working conditions of the working process, deteriorate the quality of machined surface, and reduce the tool edge life. The vibrations that occur in machine tools are called forced and self-excited vibration. The source of forced vibration in machine tools is the periodic force.

Forced vibrations are dangerous for the machine construction itself if their frequencies or higher harmonic frequencies of this force, e.g., from the cutting process, are equal to the eigen frequencies of the machine-tool-workpiece system.

If the source of the forced vibration is caused by the cutting process, the suppression of subsequent vibrations can be accomplished by selecting the cutting conditions. However, it should be borne in mind that, for example, the eigen frequencies of the workpiece can sometimes vary considerably depending on the depth of the chip being removed.

Similarly, the eigen frequency of the machine or the eigen frequency of tool clamping in the spindle may not be suitable. Another way how to suppress the forced vibration is by fixing the machine on a flexible foundation or by using a vibration absorber. On the other hand, self-excited vibrations limit the machining quality. The self-excited vibration of the machine arises without an external power supply (excitation source), since this is due to the interaction between the workpiece and the tool. If there is an excess of energy obtained, i.e., if this energy is greater than the energy consumed, self-excited vibrations occur. This is manifested as a chatter of the machine; this is caused by a number of mechanisms. Self-excited vibrations occur during roughing and finishing operations. This does not mean that if less chip is removed, self-excited vibrations are avoided. For example, self-excited vibrations may occur when removing a chip of small depth on a vertical lathe (0.3 mm) with a large load of the ram on the tool tip (1500 mm) [5].

Self-excited vibrations occur suddenly; stable conditions of cutting process can also suddenly change to unstable ones. Stable conditions become unstable when a certain value of chip depth, which is called a limit chip depth, is exceeded. The basics of the self-excited vibration theory were developed in the 1950s at VÚOSO Praha, founded by Tlustý, Poláček, and others. The theory was based on equality of energy in the feedback system. Energy is generated by the cutting process, which is the source of excitation, and consumed by vibrations (inertial mass, springs and absorbers that can replace the system) [5].

3. Types of accuracy of CNC machine tools

Under the term accuracy of machine tools, you can imagine several partial features of the machine. Accuracy will be taken differently from the perspective of the designer and from the perspective of the metrologist. From the metrological point of view, accuracy describes how close the measurement result is to the true value of the quantity. In the field of machine tools, we can talk about several types of accuracy, while the determination of accuracy is only qualitative (small, medium, and high). These are **geometric, working, and production accuracies**. Each of these accuracies has its own justification [6].

These basic three types of accuracy of CNC machine tools are complemented by other types of accuracy, namely **positioning accuracy, interpolation accuracy, volumetric accuracy, and thermal expansion**.

3.1 Geometric accuracy

Geometric accuracy describes the geometric structure of a machine tool from which the properties of functional parts affecting its working accuracy can be evaluated. It also describes the production quality of the machine and its assembly in an unloaded state. The tests are carried out on machines working under no load or under finishing conditions of machining [6].

Geometric accuracy of axes, their measurement and evaluation are given by the standard ČSN ISO 230-1. This section applies only to accuracy tests. It does not deal with the functional tests of the machine (vibrations, jerky movements of parts, etc.) or the determination of characteristic parameters (revolutions, feeds), as these tests are to be performed prior to the accuracy tests. Geometric tests consist of verifying the dimensions, shapes, and positions of components and their relative alignment. They include all operations that affect a part of the machine, such as planeness, alignment, intersection of axes, parallelism, squareness of straight lines or planar surfaces. They relate only to dimensions, shapes, positions, and relative motions that may affect the accuracy of the machine operation [7].

According to the standard, there are six geometric errors in linear (according to ČSN ISO 230 - 1) and rotary (according to ČSN ISO 230 - 7) axes, namely three translational errors—positioning error, horizontal and vertical straightness error and three angular errors. A typical three-axis CNC machine tool contains 21 geometric errors—3 × 3 translation errors, 3 × 3 angular errors. To these errors, the errors of the relative squareness of the linear axes are added. All of these errors can adversely affect the overall positioning accuracy of the machine and thus also the accuracy of the machined parts. Errors usually occur when the actual position differs from the position displayed on the machine control unit. Errors increase with dynamic effects arising from the interpolation of axes [4].

In the case of three-axis kinematics, we can find 21 error parameters, 18 translational errors and 3 parameters of squareness of individual machine axes. These errors, including spindle errors, are shown for the three-axis vertical milling machine in **Figure 1**. The kinematic chain of the three-axis machine tool presented below corresponds to W (Workpiece) -X-Y-Z-T (Tool) [8].

The error description for one linear X-axis and one rotary C-axis is given in **Table 1**.

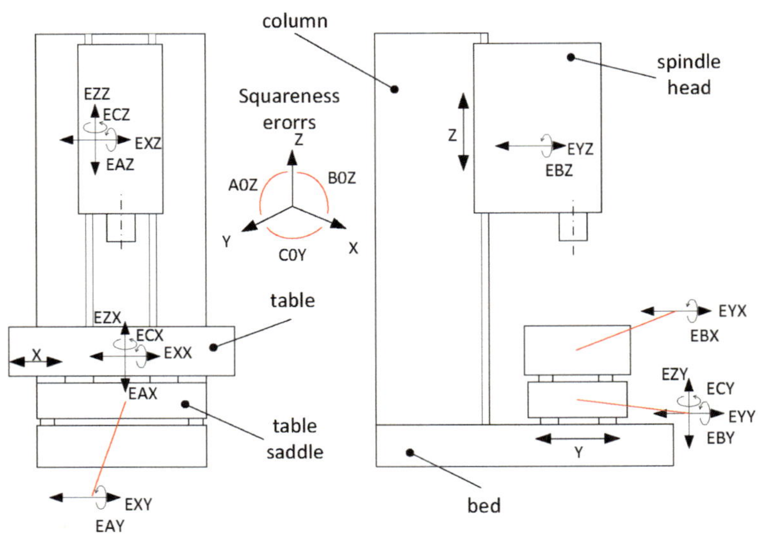

Figure 1.
Scheme of deviations of three-axis kinematics at the machine MCV 754 QUICK, KOVOSVIT-MAS [8].

Linear axis X	Rotary axis C
EXX – positioning error	EXC – radial motion in X direction
EYX – straightness error in Y direction	EYC - radial motion in Y direction
EZX - straightness error in Z direction	EZC - axial motion of C axis
EAX – angular roll error	EAC - tilt error motion around the X of the C axis
EBX - angular pitch error	EBC - tilt error motion around the Y of the C axis
ECX - angular yaw error	ECC - angular positioning error

Table 1.
Error description for one linear axis.

As early as in 1932, German professor Georg Schlesinger published a book "Inspection Test on Machine Tools," which became the basis for a unified system for assessing the accuracy of machine tools. In this book, he introduced guidelines for the use of devices and equipment for machine tool inspections. Measurement procedures and tolerances for permitted deviations are also given. The name of prof. Schlesinger is used to informally call the geometric accuracy tests of machine tools.

The devices and aids most commonly used to measure geometric errors in machine tools are, for example, granite rulers and cubes, dial gauges, digital inclinometers, autocollimators or laser interferometers, which are increasingly used for measurement. The principle of light interference as a measuring tool dates back to 1880, when Albert Michelson developed interferometry. The Michelson interferometer consists of a light source of one wavelength (monochromatic light), a silver-coated mirror and two other mirrors. Although modern interferometers are more sophisticated and measure with accuracy of the order of 1 ppm and higher, they still use the basic principles of the Michelson interferometer [4].

The straightness measurement shows deflection (bent component) or misalignment in the machine guides. This may be due to wear, an accident that may have damaged them, or poor machine foundations that cause the axis or the entire machine to drop.

Squareness is measured by comparing the straightness of two nominally orthogonal axes. Measurements can be carried out using different fixtures and devices with different arrangements. Measuring prisms, mandrels, or granite cubes may be included among fixtures while dial gauges and lasers among devices [4].

Planeness measurement is performed to check the planeness of CMM tables and machine tools, plate fields and surfaces. It determines whether there are any significant peaks or valleys and quantifies them. If these errors are significant, corrective operations are required. A certain number of measuring lines are required to measure the planeness of the surface.

3.2 Positioning accuracy

This parameter describes the accuracy and repeatability of positioning in linear and rotary numerically controlled axes. "Determination of accuracy and repeatability of positioning in numerically controlled axes" is described in the standard ISO 230-2/6 (ISO 230-2 Test code for machine tools—Determination of accuracy and repeatability of positioning numerically controlled axes; ISO 230-6 Test code for machine tools—Determination of positioning accuracy on body and face diagonals), but very often the directive VDI/DGQ 3441is also used [6].

Positioning accuracy is the most common form of measurement made with a laser interferometer (**Figure 2**). The laser system measures linear positioning accuracy and repeatability by comparing the position displayed on the machine with the actual position measured by the laser system.

A more advanced device for measurement of positioning accuracy of the machine is the Laser Tracker, which allows for immediate evaluation of the x, y, and z deviations. The geometric accuracy of the machine and the accuracy of positioning can be evaluated simultaneously (**Figure 3**) for an already assembled and activated machine. For this reason, the aforementioned accuracies are usually considered simultaneously [9].

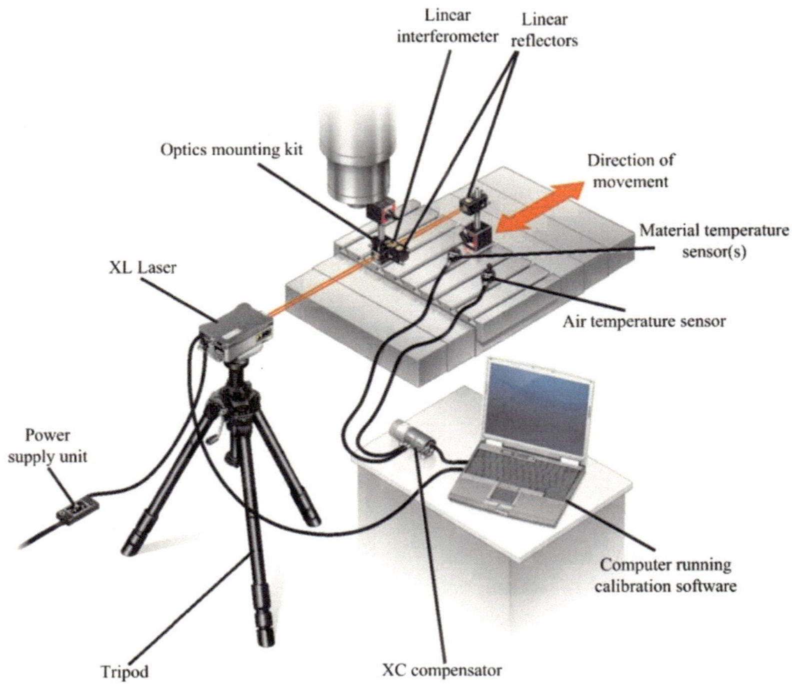

Figure 2.
Setting of measuring system for measurement of positioning accuracy [Renishaw].

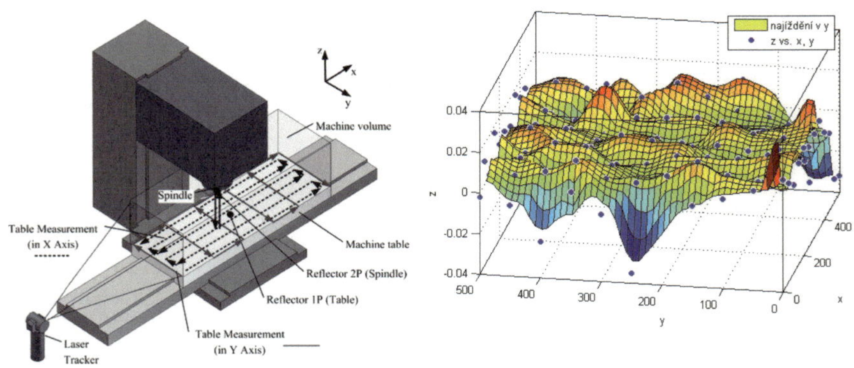

Figure 3.
Synergy when evaluating geometric and positioning accuracy using a laser tracker [9].

3.3 Interpolation accuracy

Theoretically, if the CNC machines were perfectly accurate, then the circular path of the machine would exactly match the programmed circular path. In practice, however, any of the errors (measuring error, straightness, clearance, reverse error, etc.) will cause the radius of the circle to deviate from the programmed circle. If we are able to accurately measure the actual circular path and compare it with the programmed (nominal) path, we would get a scale of the machine tool accuracy. Measurement and evaluation of circular interpolation accuracy are the subject of, for example, the standard ČSN ISO 230-4. The aim of the tests is to provide a method for estimating the properties of contour forming of numerically controlled machine tools. These errors are affected by the geometric errors and dynamic behavior of the machine at the feed used. Results are visible on machined parts under ideal machining conditions if the diameter and feed are the same for both machining and interpolation testing [1, 7].

3.4 Volumetric accuracy

Advanced and highly progressive methods include the assessment of volumetric accuracy and its subsequent compensation. The purpose of these advanced compensations is to minimize the tool center point (TCP) deviation at any point in the machine measured workspace. TCP volumetric deviation is defined as the sum of partial deviations in the individual axes [6].

Volumetric accuracy of machine tools is represented by a vector map of error deviations in the workspace. In the standard ISO 230-1, the concept of volumetric accuracy for a three-axis center is defined as the maximum range of relative deviations between the actual and ideal position in the X, Y, Z directions and the maximum range of deviations orientation for directions of A, B, C axes for motions in X, Y, Z axes in the specified volume, where the deviations are the relative deviations between the tool and the workpiece on the machine tool for specified alignment of the primary and secondary axes [1, 10].

The LaserTRACER measuring device (**Figure 4**) is mainly used for measuring of volumetric accuracy and subsequent volumetric compensation. The principle of the LaserTRACER measurement is based on measurement of beam lengths (HeNe laser wavelengths, 632.8 nm) and calculation of the measured point in the workspace by the method of sequential multilateration.

With this method, it is necessary to measure gradually from multiple locations on the machine (it is recommended to measure from at least four LaserTRACER positions). The method is presented as an analogy to the GPS system [10].

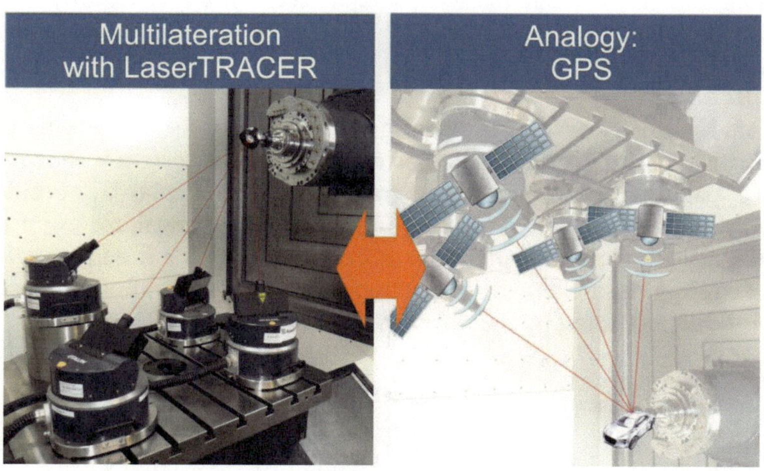

Figure 4.
Principle of measurement with LaserTRACER [etalon].

3.5 Working accuracy

This is a property of a machine tool that expresses the quality and productivity of a potential workpiece production. Working accuracy is expressed by the production of a test workpiece or a series of test workpieces. The working accuracy of the machine is affected by the accuracy of the relative tool path [6].

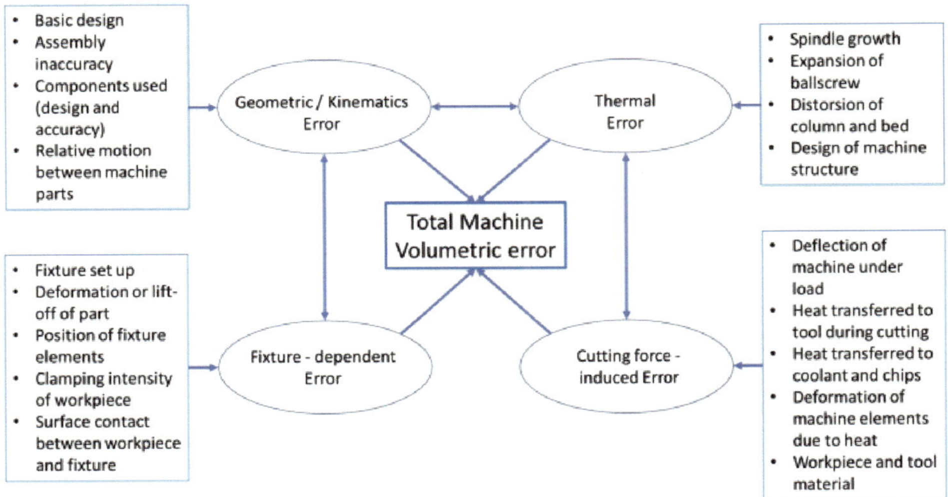

Figure 5.
Overview of the error budget in a machine tool and the factors affecting it [11].

- geometric accuracy of the machine;
- tool positioning accuracy relative to the workpiece (positioning accuracy);
- resistance of the machine to elastic deformations (caused by cutting forces, workpiece weight, etc.);
- resistance of the machine to thermal expansion ("thermal stability");
- selection of cutting conditions, etc.

An overall summary of factors affecting the accuracy of the machine tool is shown in **Figure 5**. The resulting error in the Cartesian coordinate system is shown by Eq. (1) as a spatial error between the programmed and the actual TCP position [6].

Test workpieces to be tested for working accuracy are given, for example, by ISO 10791–7. Here, a test workpiece for three-axis machining is designed. Furthermore, test workpieces are aimed at continuous five-axis machining. An example is the test workpiece defined by the directive VDI NCG 5211-1.

3.6 Production accuracy

Production accuracy describes the production process accuracy evaluated on the workpiece. Production accuracy is influenced by geometrical accuracy, positioning accuracy, working accuracy, and also by the errors of machine operator (incorrectly adjusted tool, poorly clamped workpiece) and by changes of ambient conditions. Variations in the dimensions of the test workpieces during the production process provide direct information on production accuracy [6].

Production accuracy is usually monitored by SPC (statistical process control). This method has already been overcome in some production processes with 100% product control. Due to the spectrum of workpieces of medium-sized and large CNC machine tools, the SPC method can still be considered valid [6].

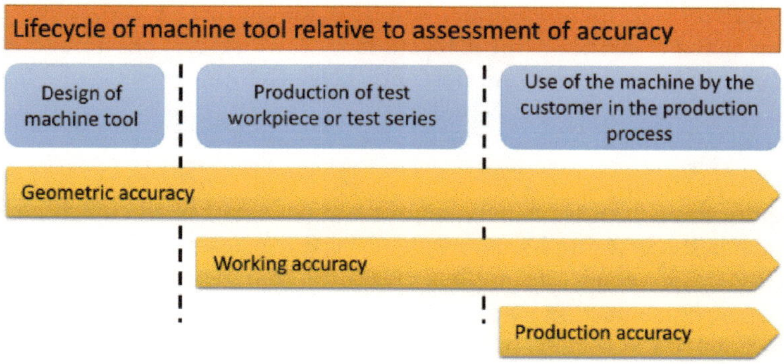

Figure 6.
Relationships between individual accuracies of a CNC machine tool throughout its life cycle.

The three main influences that affect the machine tool and the production process and cause workpiece dimensional variations can be more closely assigned to [4]:

- production technology 15%,
- working accuracy of the machine 25%,
- measurement 15%,
- ambient conditions 20%,
- machined part 5%,
- machine operator 20%.

The above-mentioned partial accuracies of the machine tool can be divided into individual parts of the life cycle (**Figure 6**). Production accuracy can, therefore, be monitored at the phase of customer's machine use and is influenced by both the working accuracy of the machine and long-term stability of geometric accuracy.

4. Example of basic compensation: positioning accuracy

One of the possibilities of compensating the error of linear and rotary axis is to use the so-called interpolation compensations, which include the compensation of leadscrew errors and measuring system errors [12]. In the SIEMENS control system, errors are referred to as LEC and MSEC (*LEC*-Leadscrew Error Compensation and *MSEC*-Measuring System Error Compensation). Compensation values are entered into the system in the form of tables, which are either manually entered or subprograms can be generated using various software solutions, which automatically load and write the table into the machine control system. Here, it should be taken into account that this automatic table loading can only work for given versions of the machine control system with the appropriate service pack. The MSEC compensation is also referred to as ENC_COMP in the machine control system and, through this parameter, the compensation is gradually set and activated. The abbreviations depend on the type of machine control system.

Only unidirectional compensations can be made by ENC_COMP compensation. In the event that a clearance error is found from the test, it is possible to use the Backlash compensation in combination with ENC_COMP.

4.1 Backlash

During the transfer of force between the movable part of the machine and its drive—e.g., a ball screw and its mounting—there are clearances (gaps) at different load directions. Conversely, a complete clearance-free mechanical adjustment will dramatically increase machine wear and heat generation. Mechanical clearances cause deviations in the reverse path of axes or spindles with indirect measuring systems. This means that if the direction changes, the axis will travel depending on the gap size. These clearances are compensated by the function listed below as Backlash.

Backlash can be entered into the control system in several ways. The first option is to use the machine parameter and enter the value as a constant for the selected axis.

The second option is to use the SAG compensations and the CEC table, which will be described in the next step and eliminate the clearance error by bidirectional compensation. The advantage of the first solution is to specify only one constant. In the case of non-linear behavior, it is preferable to enter the clearance in the form of a CEC table.

To use the MSEC compensation, the table for the Siemens control system will be as follows:

```
%_N_AX_EEC_INI
CHANDATA(1)
$AA_ENC_COMP[0,0,X1]=0.003     ; first compensation value (interpolation point
                                 0):+3µm
$AA_ENC_COMP[0,1,X1]=0.01      ; second compensation value (interpolation
                                 point 1): +10µm
$AA_ENC_COMP[0,2,X1]=0.012     ; third compensation value (interpolation
                                 point 2): +12µm

$AA_ENC_COMP[0,800,X1]=-0.0    ; last compensation value (interpolation point
                                 800): 0µm
$AA_ENC_COMP_STEP[0,X1]=1.0    ; Distance between two compensation values
                                 1.0 mm
$AA_ENC_COMP_MIN[0,X1]=-200.0 ; Start of compensation -200.0 mm
$AA_ENC_COMP_MAX[0,X1]=600.0 ; End of compensation +600.0 mm
$AA_ENC_COMP_IS_MODULO[0,X1]=0 ; Compensation without modulo
M17                              function
```

5. Example of basic compensation: sag compensation

In the previous paragraph, compensation in one MSEC axis was described [12]. In a large number of cases, MSEC compensation is insufficient and it is advisable to introduce corrections of two dependent axes. The sag compensation is performed when the weight of the individual machine elements leads to the positioning displacement and inclination of the moving parts, as this causes the related machine parts—including guide systems—to bend. The compensation error of angle is used

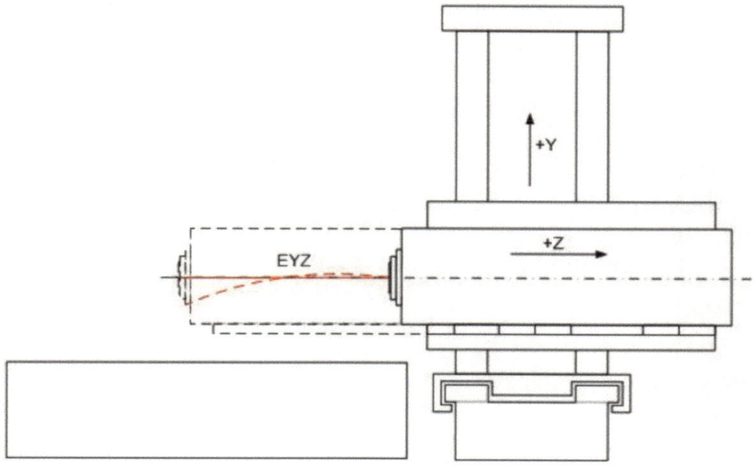

Figure 7.
Error EYZ of horizontal machine tool.

when the motion axes are not properly aligned at the correct angle (e.g., vertical). As the deviation from the zero position increases, the positioning errors also increase. Both types of errors can occur as a result of shifting the weights of individual machine parts, replaceable heads, workpiece diversity, and machine compliance. Measured correction values are calculated based on the relevant standards or own algorithms and are stored in the machine control system in the form of a compensation table during commissioning.

During machine operation and motion of axes, the corresponding value is interpolated between the values of the "interpolation points" table. For each motion in a continuous path, there is always both the base axis and the compensation axis. If the perpendicular y-axis is not in the continuous path of the x-axis and the y-axis, this inaccuracy is compensated by the x-axis in the continuous path. **Figure 7** shows the principle of compensation on an example of a horizontal machine tool. The straightness error of EYZ is largely due to the machine compliance, while, through the ram travel, the sag occurs which is caused by the load of the assembly spindle-ram-slide-accessory.

This compensation provides a wide range of options for elimination of geometric errors. Here, an example will be given to compensate a sag, e.g., caused by changing the load of the replaceable heads, where there may be significant differences in their weights. If the machine is without a replaceable head, the sag is shown in **Figure 7**. If a milling head with a certain weight is used, the travel will be more loaded; therefore, a greater deformation will occur.

To use the SAG compensation for sagging compensations, the table for the Siemens control system will be as follows:

```
%_N_NC_CEC_INI
CHANDATA(1) ;
$AN_CEC[0,0]=0       ; first compensation value (interpolation point 0); for Z:
                       ±0μm
$AN_CEC[0,1]=0.01    ; second compensation value (interpolation point 1); for Z:
                       +10μm
$AN_CEC[0,2]=0.012   ; third compensation value (interpolation point 2); for Z:
                       +12μm
```

```
$AN_CEC[0,100]=0    ; last compensation value (interpolation point 101); for Z:
                      ±0µm
$AN_CEC_INPUT_AXIS[0]=(AX2)    ; base axis Y
$AN_CEC_OUTPUT_AXIS[0]=(AX3) ; compensation in Z axis
 $AN_CEC_STEP[0]=8        ; distance between interpolation points 8.0 mm
$AN_CEC_MIN[0]=0          ; start of compensation Y=0 mm
$AN_CEC_MAX[0]=800.0         ; end of compensation Y=800 mm
$AN_CEC_DIRECTION[0]=0    ; table applies to both directions of Y axis motions
$AN_CEC_MULT_BY_TABLE[0]=0;
$AN_CEC_IS_MODULO[0]=0    ; compensation without modulo function
M17 ;
```

If we use the SAG compensations for bidirectional axis compensation, the table for the Siemens control system will be as follows. The parameters of both the base axis and the compensated axis will be the same and match the axis designation. The direction parameter will be first set to 1 and then to −1. As an example of a horizontal boring machine, for the Z axis of ram travel, it will be as follows.

```
%_N_NC_CEC_INI ;
CHANDATA(1) ;
$AN_CEC[0,0]=0    ; first compensation value (interpolation point 0); for Z: ±0µm
$AN_CEC[0,1]=0.01 ; second compensation value (interpolation point 1); for Z:
                    +10µm
$AN_CEC[0,2]=0.012    ; third compensation value (interpolation point 2); for Z:
                        +12µm
$AN_CEC[0,10]=0       ; last compensation value (interpolation point 11); for Z:
                        ±0µm
$AN_CEC_INPUT_AXIS[0]=(AX3)    ; base axis Z
$AN_CEC_OUTPUT_AXIS[0]=(AX3) ; compensation in Z axis
$AN_CEC_STEP[0]=75         ; distance between interpolation points 75.0 mm
$AN_CEC_MIN[0]=0.0         ; start of compensation in Z=0 mm
$AN_CEC_MAX[0]=750.0    ; end of compensation in Z=750 mm
$AN_CEC_DIRECTION[0]=1    ; table applies to only positive direction of Z axis
$AN_CEC_MULT_BY_TABLE[0]=0    ;
$AN_CEC_IS_MODULO[0]=0            ; compensation without modulo function
$AN_CEC[0,0]=0    ; first compensation value (interpolation point 0); for Z: ±0µm
$AN_CEC[0,1]=0.01    ; second compensation value (interpolation point 1); for Z:
                       +10µm
$AN_CEC[0,2]=0.012    ; third compensation value (interpolation point 2); for Z:
                        +12µm
$AN_CEC[0,11]=0    ; last compensation value (interpolation point 11); for Z:
                     ±0µm
$AN_CEC_INPUT_AXIS[0]=(AX3)    ; base axis Z
$AN_CEC_OUTPUT_AXIS[0]=(AX3) ; compensation in Z axis
$AN_CEC_STEP[0]=75    ; distance between interpolation points 75.0 mm
$AN_CEC_MIN[0]=0.0    ; start of compensation in Z=0 mm
$AN_CEC_MAX[0]=750.0    ; end of compensation in Z=750 mm
$AN_CEC_DIRECTION[0]=-1    ; table applies to only positive direction of Z axis
$AN_CEC_MULT_BY_TABLE[0]=0    ;
$AN_CEC_IS_MODULO[0]=0    ; compensation without modulo function
M17 ;
```

Furthermore, SAG compensations are used to compensate squareness error. The squareness compensations of the Siemens control system are entered using CEC tables, where one axis is determined as the base axis and the other as compensated. An example will be given to compensate the squareness of, for example, the Y and Z axes of a horizontal machining center. From the measured values obtained, for example, from measurements with a laser interferometer, ballbar or calibration cubes and dial gauges, we obtain information on the size and orientation of squareness, which may be, for example, 22.4 μm/m. It is necessary to respect the machine coordinate system and orientation of axes when preparing the measurements. Otherwise, for the verification measurement, the resulting error value will be multiplied. For a ram travel (Z axis), this means that for a travel length of 750 mm, the measured error of 22.4 μm/m must first be converted by a ratio of 750/1000 mm. After multiplying by the measured value, we obtain the value for entering the correction into the machine control system. In this case, the value at the 750 mm position will be 16.8 μm.

For the above example, the compensation table for travel of the ram axis Z will be as follows.

```
%_N_NC_CEC_INI ;
CHANDATA(1) ;
$AN_CEC[0,0]=0    ; first compensation value (interpolation point 0); for Z: ±0μm
$AN_CEC[0,100]=0.0168 ; last compensation value (interpolation point 11); for Z:
                        ±16.8μm
 $AN_CEC_INPUT_AXIS[0]=(AX3)    ; base axis Z
$AN_CEC_OUTPUT_AXIS[0]=(AX2) ; compensation in Y axis
$AN_CEC_STEP[0]=750        ; distance between interpolation points 750.0
$AN_CEC_MIN[0]=0.0        ; start of compensation in Z=0 mm
$AN_CEC_MAX[0]=750.0        ; end of compensation in Z=750 mm
$AN_CEC_DIRECTION[0]=0   ; table applies to for both directions of Z axis
                            motions
$AN_CEC_MULT_BY_TABLE[0]=0    ;
$AN_CEC_IS_MODULO[0]=0    ; compensation without modulo function
M17 ;
```

6. Example of advanced compensation: volumetric compensation

The DMU 75 monoBlock® machine (**Figure 8**) is kinematically adapted to have three linear motions in the tool (X = 750, Y = 650, Z = 560 mm) and two rotary motions in the workpiece (swinging about the X axis and rotation around the Z axis). It is equipped with the Heidenhain TNC 640 control system. This machine has a positioning accuracy of 8 μm per axis.

The measurement and compensation of the volumetric accuracy of the linear machine axes are shown in **Figure 9**. After compensation, the workspace was improved by approx. 60%.

Before verification measurement of the volumetric accuracy, the machine was measured by a DBB device to verify the successful activation of volumetric compensation. **Figure 10** shows an improvement in the accuracy of circular interpolation on the shape of roundness (especially squareness); therefore, the machine was verified by the LaserTRACER to detect an improvement in overall volumetric accuracy [13].

After compensating the volumetric accuracy of the linear axes, the rotary axis that is the first in the kinematic chain from the workpiece to the tool, i.e., the C axis, must first be measured. This axis was measured with an example of the results in **Figures 11** and **12** [13].

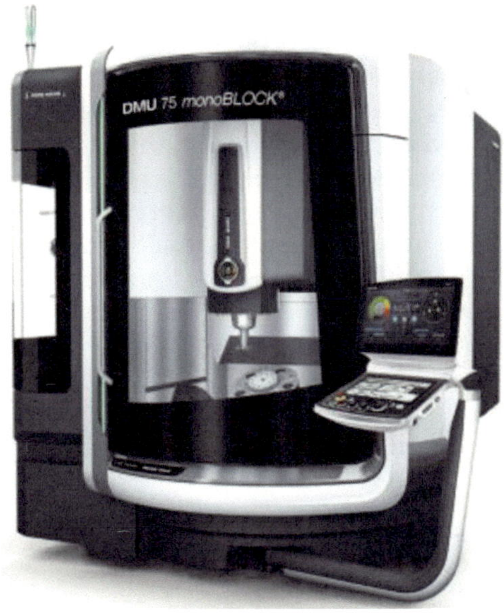

Figure 8.
View of DMU 75 monoBlock ® [DMG Mori].

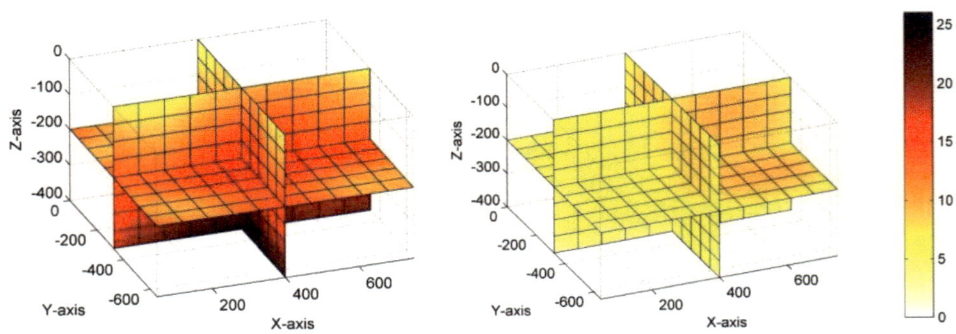

Figure 9.
Results of volumetric accuracy measurement of linear axes before and after compensation [13].

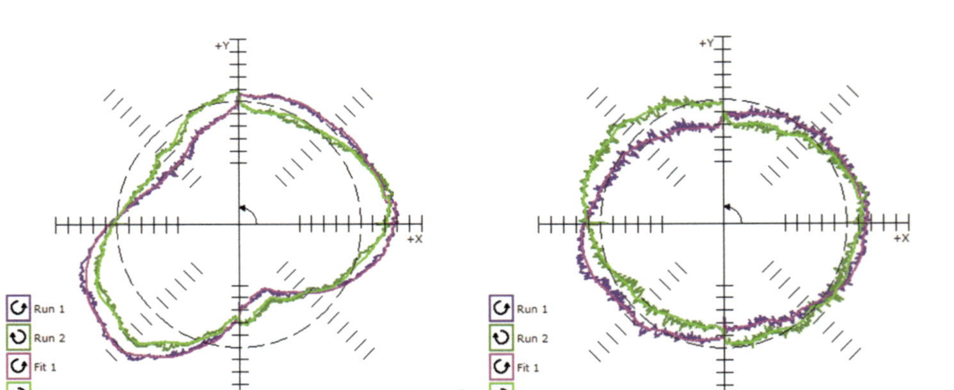

Figure 10.
Accuracy of circular interpolation in XY plane before and after volumetric compensation [13].

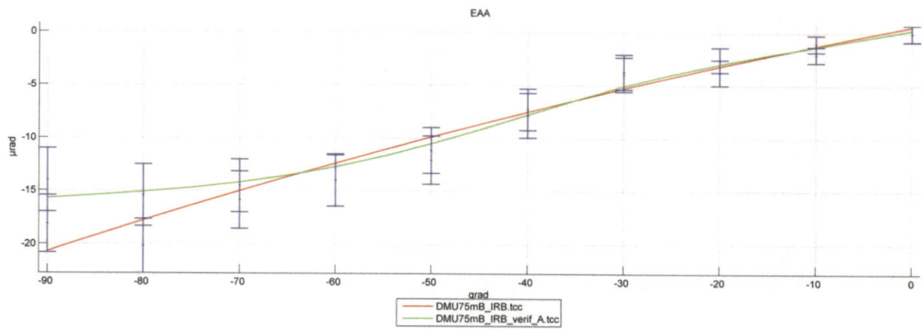

Figure 11.
Error of EAA axis A [13].

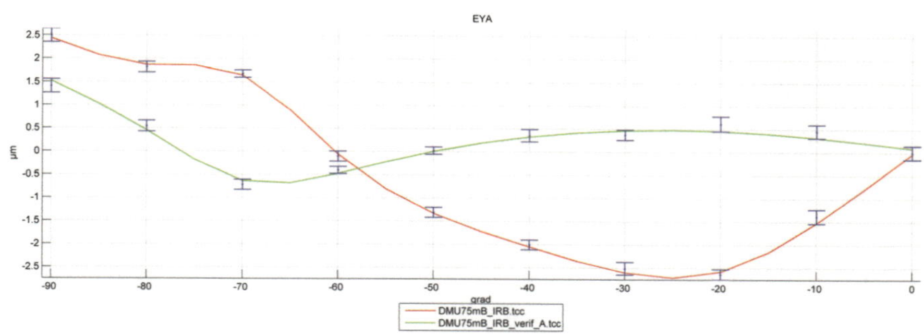

Figure 12.
Error of EYA axis A [13].

7. Conclusion

The aforementioned accuracies are related to one another and it cannot be assumed, for example, that the desired working accuracy can be achieved by poor geometric accuracy. **Figure 13** shows cascading of these accuracies.

Figure 13 shows a machine tool with linear axes. If there are rotary axes on the machine, it is necessary to check the linear axes first and then check the rotary axes. These are also checked for geometrical, positioning, and volumetric accuracy. If all

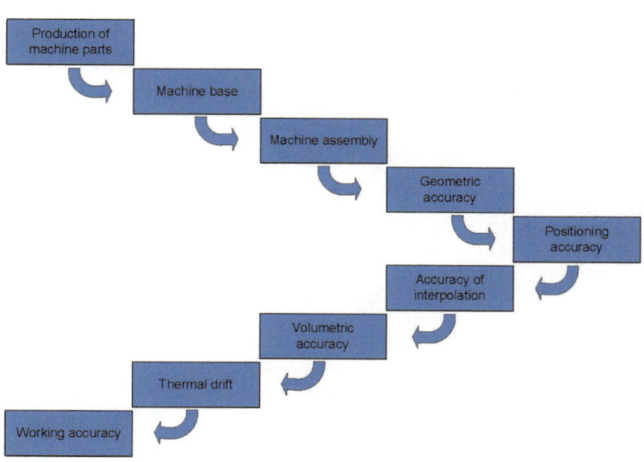

Figure 13.
Cascading of accuracies in machine tools [13].

the accuracies are within the required tolerances, the working accuracy related to the machining of the workpiece can be stepped to. Individual accuracies are described in the following section.

Acknowledgements

These results were obtained with the financial support of the Faculty of Mechanical Engineering, Brno University of Technology (Grant No. FSI-S-20-6335).

Author details

Jiri Marek*, Michal Holub, Tomas Marek and Petr Blecha
Institute of Production Machines, Systems and Robotics, Brno University of Technology, Brno, Czech Republic

*Address all correspondence to: marek@fme.vutbr.cz

References

[1] Marek T. Měření a kontrola obráběcích strojů. In: Konstrukce CNC Obráběcích Strojů IV. Praha: MM Publishing, s.r.o.; 2018. pp. 336-350. ISBN: 978-80-906310-8-3

[2] Marek J. Konstrukce CNC Obráběcích Strojů III. Praha: MM Publishing, s.r.o.; 2014. ISBN: 978-80-260-6780-1

[3] Holub M, Andrs O, Kovar J, Vetiska J. Effect of position of temperature sensors on the resulting volumetric accuracy of the machine tool. Measurement [Online]. 2020;**150**: 107074. DOI: 10.1016/j.measurement. 2019.107074. ISSN: 02632241

[4] Marek T, Marek J. Mít Sondu Nestačí. Brno: Renishaw s.r.o.; 2017. ISBN: 978-80-87017-20-3

[5] Marek J. Stavba CNC Obráběcích Strojů—Souvislosti a Fakta. Praha: Grumant s.r.o.; 2017

[6] Holub M. Geometric accuracy of machine tools. In: Measurement in Machining and Tribology [Online]. Materials Forming, Machining and Tribology. Cham: Springer International Publishing; 2019. pp. 89-112. DOI: 10.1007/978-3-030-03822-9. ISBN: 978-3-030-03821-2

[7] ČSN ISO 230-1 Zásady Zkoušek Obráběcích Strojů - Část 1: Geometrická Přesnost Strojů Pracujících bez Zatížení Nebo za Dokončovacích Podmínek Obrábění. Praha: Úřad pro Technickou Normalizaci, Metrologii a Státní Zkušebnictví; 2014

[8] Holub M, Blecha P, Bradac F, Kana R. Volumetric compensation of threeaxis vertical machining centre. MM Science Journal [Online]. 2015;**2015**(03): 677-681. DOI: 10.17973/MMSJ.2015_10_ 201534. ISSN: 18031269

[9] Knobloch J, Holub M, Kolouch M. Laser tracker measurement for prediction of workpiece geometric accuracy. In: Engineering Mechanics. Vol. 1. Svratka: Institute of Solid Mechanics, Mechatronics and Biomechanics; 2014. pp. 296-300. ISBN: 978-80-214-4871-1. ISSN: 1805-8248

[10] Holub M, Kol. GTS—Testování Obráběcích Strojů. Brno: VUT Brno – UVSSR; 2016

[11] Ramesh R, Mannan MA, Poo AN. Error compensation in machine tools— A review. International Journal of Machine Tools and Manufacture [Online]. 2000;**40**(9):1235-1256. DOI: 10.1016/S0890-6955(00)00009-2. ISSN: 08906955

[12] Holub M. Kompenzace geometrické přesnosti CNC obráběcích strojů. In: Konstrukce CNC Obráběcích Strojů IV. Praha: MM Publishing; 2018. pp. 352-364. ISBN: 978-80-906310-8-3

[13] Marek T. Predikování Vybraných Vlastností Rotačních Kinematických Dvojic Obráběcích Strojů [Online]. Vysoké Učení Technické v Brně. 2019. Available from: https://www.vutbr.cz/ studenti/zav-prace/detail/122509

Chapter 6

Actively Controlled Journal Bearings for Machine Tools

Jiří Tůma, Jiří Šimek, Miroslav Mahdal, Jaromír Škuta, Renata Wagnerová and Stanislav Žiaran

Abstract

The advantage of journal hydrodynamic bearings is high radial load capacity and operation at high speeds. The disadvantage is the excitation of vibrations, called an oil whirl, after crossing a certain threshold of the rotational speed which depends on the radial bearing clearance, the viscosity of lubricating oil, and the rotor mass. A passive way of how to suppress vibrations consists of adjusting the shape of the bearing bushing. Vibrations can be suppressed using the system of active vibration damping with piezoactuators to move the bearing bushing in two directions. The displacement of the bearing bushing is actuated by two piezoactuators, which respond to the position of the bearing journal relative to the bearing housing. Two stacked linear piezoactuators are used to actuate the location of the bearing bushing. A pair of capacitive sensors senses the position of the journal or shaft. The system of the actively controlled journal bearings is the first functional prototype in the known up to now. It works with a cylindrical bushing which does not require special technology of manufacturing and assembly. This new bearing enables not only to damp vibrations but also serves to maintain the desired bearing journal position with an accuracy of micrometers.

Keywords: journal bearings, active vibration control, piezoactuators, oil whirl

1. Introduction

The reasons for the interest in actively controlled journal bearings (alternatively called sleeve bearings or plain bearings for radial load) lie in demand for the introduction of high-speed cutting or machining as a technology for the future. Increasing the machining speed was required to get beyond the limits of the interval, where unwanted temperature increases. Some researchers define high-speed machining as machining whereby conventional cutting speeds are exceeded by a factor of 5–10. Increased machining speed has advantages. The ability to benefit the advantages of high-speed cutting in steel, cast iron, and nickel-based alloys can be obtained with spindle speeds in the range of 8k to 12k rpm. High-speed cutting of nonferrous materials such as brass, aluminum, and engineered plastics demands a significantly higher rpm capability. For these materials, we must focus on milling equipment capable of operating at high-speed spindle speeds of 25k to 50k rpm or more. High-speed machining can also include grinding and turning.

Let us now notice the machine tool spindles. Roller bearings support these spindles. Prestressed ball or tapered roller bearings are used to eliminate play. This chapter focuses on plain radial bearings, namely, hydrodynamic bearings. Plain bearings of this type require a clearance for their function, which is selected in the range of 0.1–0.3% of the journal diameter.

Journal hydrodynamic bearings are a standard solution to support rotors. Their advantage is a possibility to carry the high radial load and to operate at high rotational speeds. The disadvantage of the journal bearings is the excitation of unwanted rotor vibrations by whirling of the journal in the bearing bushing. The bearing journal becomes unstable as the journal axis begins to perform a circular motion that is bounded only by the walls of the bearing bushing. When the speed threshold is exceeded, the axis of the bearing journal starts to circulate, causing the rotor to vibrate. These vibrations are called whirl. A passive way of how to suppress vibrations consists in adjusting the shape of the bearing bushing, such as lemon or elliptical bore of the bushing, or the use of tilting pads. Even though there are several solutions based on mentioned passive improvements, this article deals with the use of active vibration control (AVC) with piezo-actuators as a measure to prevent instability.

The disadvantage of bearings of this type is whirl instability, which can cause machine tool vibrations. The following chapter describes the possible operating range of the spindle speed.

2. Operating speed range of plain bearings

Special oil for high-speed spindle bearing of the OL-P03 type was used for testing (VG 10 grade, viscosity μ = 0.027 Pa.s at 20°C). Tests were carried out without preheating the lubricant at a normal temperature. The oil viscosity at ambient temperature in the laboratory corresponded to the oil viscosity at 40°C in industrial bearings. The journal bearing cross-section is shown in **Figure 1**.

The operating conditions of the hydrodynamic bearing are described by the Sommerfeld number [1]:

$$S = (R/c)^2 \mu N/P, \tag{1}$$

where N is a rotational speed of the rotor in rev/s, μ is a dynamic viscosity in Pa.s, R is a radius of the journal, c is a radial clearance, P is a load per unit of projected bearing area ($2RL$) in N/m^2, where L is a bearing length.

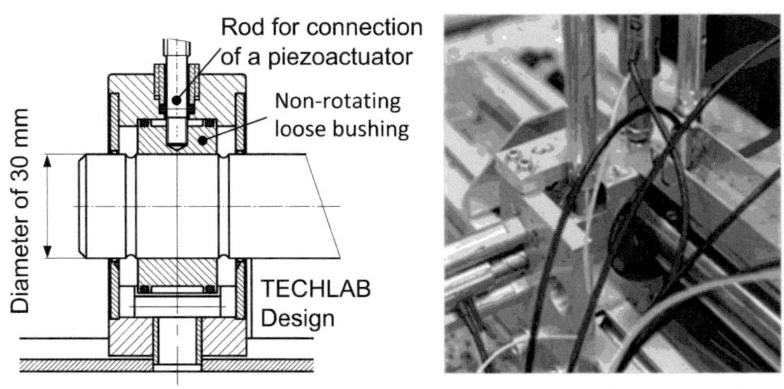

Figure 1.
Side-view technical drawing and a photo of bearing housing.

The value of the Sommerfeld number for the given bearing size and the rotor mass of 0.83 kg is as follows: $R = 0.014 \times N$, where N is the mentioned rotational speed.

The magnitude of friction coefficient in the plain bearings was analysed in the past by the McKee brothers [2]. It has been found that bearing friction is dependent on a dimensionless bearing characteristic given by a ratio $\mu N/P$ whose parameters are defined above. If the rotor does not rotate or rotate slowly, there is only a very thin film between the journal and the bushing. Boundary or thin-film boundary lubrication occurs with a considerably increased coefficient of friction. Many experiments show that the journal axle moves chaotically at low speeds or the journal starts to oscillate. It is uncertain at the start of run-up whether the axle of the bearing journal moves to the left or right, regardless of the direction of rotation. Only when the specified speed limit is exceeded the lubrication becomes hydrodynamic, and thick film of the lubricant is formed, and the trajectory of motion can be predicted. The limit value of the bearing characteristic for the boundary lubrication is described in [1]. Designers keep the value $\mu N/P \geq 1.7 \times 10^{-6}$ (reyn × rev/s/psi), which is about five times the value the McKee brothers have determined. The measurement in our test rig shows the limit of the boundary lubrication at about 1k rpm, which corresponds to the value of the dimensionless characteristic $\mu N/P$ equaled to 3.8×10^{-5} (Pa.s × rev/s/Pa) when using SI units for the input parameters. Our estimate for the lower limit of hydrodynamic lubrication corresponds to the recommendations in the handbook [1]. In experiments with the active vibration control, the feedback is closed only for stable lubrication.

An example of a gradual change of position of the bearing journal centre during an increase in speed up to 7k rpm at the constant increase rate is shown in **Figure 2**. The lubrication is of the boundary type in the range up to about 1.2k rpm and is accompanied by oscillations.

The reason for the oscillations is the step change of speed to about 300 rpm after switching on because it is not possible to increase the rotational speed continuously from zero. Hydrodynamic lubrication at stable motion is produced for rotational speed up to 5k rpm. Motion instability of the whirl type occurs when this speed of 5k rpm is exceeded. Fluid force makes sense to be modeled just for stable motion

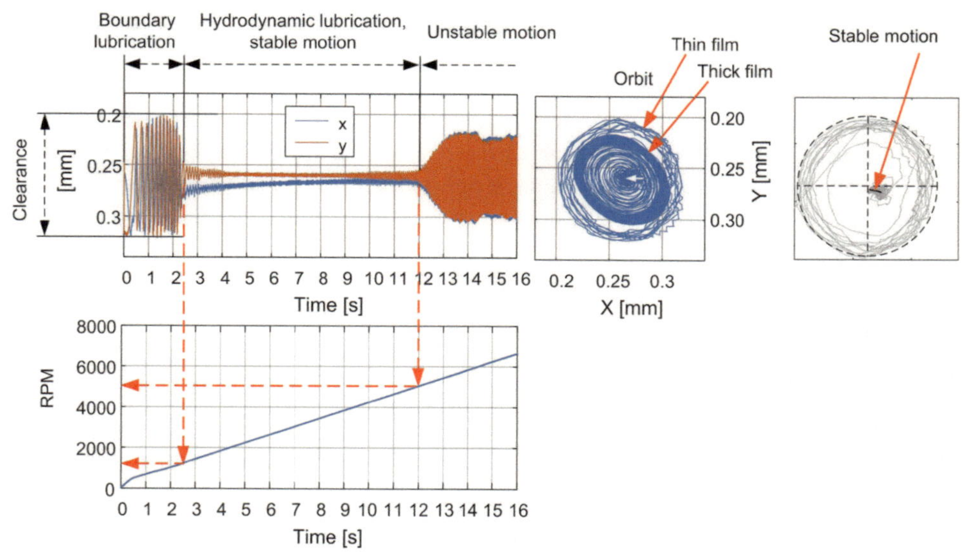

Figure 2.
The run-up of a journal bearing.

and hydrodynamic lubrication. It is almost impossible to determine the initial conditions for boundary lubrication. Notice that the centre of the journal rises to the level of the centre of the bearing bore and gradually approaches this centre so that the small eccentricity gradually decreases to zero as is shown on the right panel of **Figure 2**. The data for this orbit was approximated by the five-degree polynomial in the time interval which begins at the 3rd second and ends at the 12th second. The difference between thin and thick film lubrication is also evident on the right panel of **Figure 2**, which depicts an orbit plot for the entire measurement time up to sixteenth-second.

The threshold of the instability of the journal movement in the bearing is given by the clearance and viscosity of the oil, which depends on the temperature.

3. Instrumentations

For developing a new design of the actively controlled bearing, a test rig was built; see **Figure 3**. This figure provides different views of the test rig. An inductive motor of 400 Hz drives the rotor, and therefore the maximum rotational speed is 23k rpm. The engine is connected to the rotor via the Huco diaphragm coupling. The bearing diameter is 30 mm, and the length-to-diameter ratio is equal to about 0.77. The span of bearing pedestals is of 200 mm. The results of the experiments presented in this article are for the radial clearance of 45 μm. Also, the journals of the other clearance are available for testing. The performance of the actively controlled bearing was tested on the test bench (Rotorkit) of the TECHLAB design [3, 4]. Additionally, it should be emphasised that research was focused at rigid rotors and the journal bushing of the cylindrical bore, where the journal motion is measured at the location closest to the bearing bushing. The research work resulted in putting into operation of the active vibration control system, which became the first functional bearing prototype known up to now [5].

The mechanical arrangement of the actively controlled bearing is shown on the right of **Figure 1**. Oil leakage from the volume between the bearing body and the loose bushing and the piezoactuator rod is sealed with rubber O-rings. As it was stated before, vibrations of the rotor is suppressed using the system for active vibration control with piezoelectric actuators enabling to move the non-rotating loose bushing. The motion of the bearing bushing is controlled by the controller, which responds to the change in position of the bearing journal related to the bearing housing. Two stacked linear piezoactuators are used to actuate the position of the bearing journal via the position of the bearing bushing. The bearing uses a cylindrical bushing which did not require unique technology of production and assembly. This new bearing enables not only to damp vibrations and to prevent

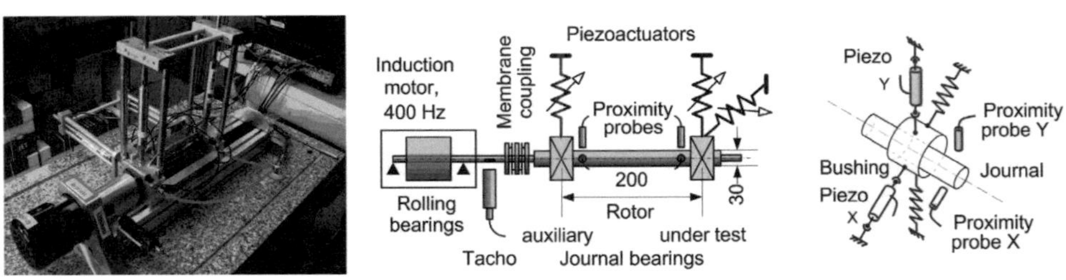

Figure 3.
Actively controlled journal bearings.

instability but also enables to maintain the desired bearing journal position with an accuracy of micrometers.

4. Mathematical model

4.1 Equation of motion

The bearing journal can be considered as a rigid body rotating within the bearing housing at an angular velocity Ω. For simplicity, it is assumed that the rotation axis does not change its direction in contrast to a model [6]. Fluid forces are caused by the hydrodynamic pressure generated in the oil film, whose total mass relative to the journal and rotor is negligible. The oil pumped by the rotating journal surface produces an oil wedge that lifts up the bearing journal so that it does not touch the inner walls of the housing. The coordinate system of a cylindrical journal bearing is shown on the left side in **Figure 4**. The planar motion of the bearing journal at the x and y coordinates can be described by two motion equations arranged into a matrix equation:

$$\begin{bmatrix} M & 0 \\ 0 & M \end{bmatrix}\begin{bmatrix} \ddot{x} \\ \ddot{y} \end{bmatrix} + \begin{bmatrix} B_{XX} & B_{XY} \\ B_{YX} & B_{YY} \end{bmatrix}\begin{bmatrix} \dot{x} \\ \dot{y} \end{bmatrix} + \begin{bmatrix} C_{XX} & C_{XY} \\ C_{YX} & C_{YY} \end{bmatrix}\begin{bmatrix} x \\ y \end{bmatrix} = \begin{bmatrix} F_X \\ F_Y \end{bmatrix}, \tag{2}$$

where M is a mass of the rotor; F_X is a force acting on the journal in the horizontal direction; F_Y is a force acting on the journal in the vertical direction; C_{UV} is a stiffness coefficient; and B_{UV} is a damping coefficient, where U is equal to X or Y and V is equal to X or Y as well.

In addition to force components in the horizontal and vertical directions, the force balance will be solved in other possible directions. Force in the direction of the line of the centers is denoted as a direct force F_D, while force which is perpendicular to the line of centers is denoted as a quadrature force F_Q. Both these forces balance the gravity force G, as is shown in the right panel of **Figure 4**.

The system is described by two motion equations, and therefore the total order of the system is four. This system may become unstable even for positive parameter values such as stiffness and damping.

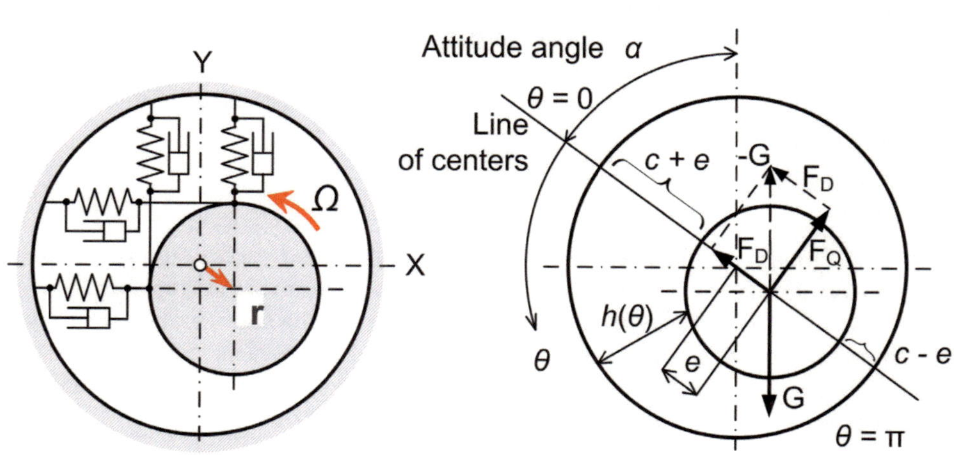

Figure 4.
A cross-section of the hydrodynamic bearing.

4.2 Muszynska model

The motion equation of the rotor with the journal bearing in coordinates x and y was designed by Muszynska. The derivation is based on the design of the formula to calculate the already mentioned direct and quadrature forces. Compared to Eq. (2), the stiffness and damping matrices are designed in such a way that the oil film is replaced by a spring and a dashpot system that rotates at an angular velocity Ω, where λ is a dimensionless parameter, which is slightly less than 0.5. The stiffness of the spring is designated by K, and the damper has a damping factor D:

$$\begin{bmatrix} M & 0 \\ 0 & M \end{bmatrix}\begin{bmatrix} \ddot{x} \\ \ddot{y} \end{bmatrix} + \begin{bmatrix} D & 0 \\ 0 & D \end{bmatrix}\begin{bmatrix} \dot{x} \\ \dot{y} \end{bmatrix} + \begin{bmatrix} K & D\lambda\Omega \\ -D\lambda\Omega & K \end{bmatrix}\begin{bmatrix} x \\ y \end{bmatrix} = \begin{bmatrix} F_X \\ F_Y \end{bmatrix}. \quad (3)$$

The derivation of the motion equation is described, for example, in the article [5].

4.3 Analytical solution of the Reynolds equation

The theory of hydrodynamic bearing is based on a differential equation derived by Osborne Reynolds. Reynolds equation is based on the following assumptions: The lubricant obeys Newton's law of viscosity and is incompressible. The inertia forces of the oil film are negligible. The viscosity μ of the lubricant is constant, and there is a continuous supply of lubricant. The effect of the curvature of the film concerning film thickness is neglected. It is assumed that the film is so thin that the pressure is constant across the film thickness. The shaft and bearing are rigid.

Furthermore, it is assumed that the thickness h of the oil film depends on the other two coordinates, namely, the coordinate z along the axis of rotation and the location on the perimeter of the journal which is described by the angle θ as is shown on the right side in **Figure 4**. If the radius of the bearing journal is equal to R, then the most general version of the Reynolds equation for calculation of the oil pressure distribution $p(\theta, z)$ is as follows (Dwivedy) [7]:

$$\frac{1}{R^2}\frac{\partial}{\partial\theta}\left(h^3\frac{\partial p}{\partial\theta}\right) + \frac{\partial}{\partial z}\left(h^3\frac{\partial p}{\partial z}\right) = 6\mu\Omega\frac{\partial h}{\partial\theta} + 12\mu\frac{\partial h}{\partial t}. \quad (4)$$

There is no analytical solution for the Reynolds equation.

During operation, the journal axis shifts from the centre of the bearing bushing to the distance of e, called eccentricity, which is related to a radial clearance c. Variable is called an eccentricity ratio $n = e/c$. The film thickness as a function of θ is defined as follows:

$$h = c(1 + n\cos\theta). \quad (5)$$

The oil film moves in adjacent parallel layers at different speeds, and shear stress results between them. The oil layer at the surface of the journal moves at the peripheral velocity of the journal, while the oil layers at the surface of the bearing bushing do not move (at zero velocity). The surface of the journal moves at a velocity of $U = R\Omega$ in m/s. Reynolds equation will be solved for the steady state and independence of the pressure distribution on the coordinate of z:

$$\frac{d}{d\theta}\left[h^3\frac{dp}{d\theta}\right] = 6\mu UR\frac{dh}{d\theta}. \quad (6)$$

On double integrating, see [7], we get

$$\begin{aligned} h^3\frac{dp}{d\theta} &= 6\mu UR\int\frac{dh}{d\theta}d\theta = 6\mu URh + K \\ \frac{dp}{d\theta} &= 6\mu UR\left[\frac{1}{(1+n\cos\theta)^2}+\frac{K_1}{c(1+n\cos\theta)^3}\right] \\ p(\theta) &= \frac{6\mu UR}{c^2}\int\left[\frac{d\theta}{(1+n\cos\theta)^2}+\frac{K_1 d\theta}{c(1+n\cos\theta)^3}\right]+p_0, \end{aligned} \tag{7}$$

where K, K_1, and p_0 are integration constants.
The solution must meet the boundary condition:

$$p(\theta=0)=p(\theta=2\pi)\Rightarrow p(\theta=0)-p(\theta=2\pi)=0, \tag{8}$$

which gives

$$\begin{aligned} &\frac{6\mu UR}{c^2}\int_{\theta=0}^{\theta=2\pi}\left[\frac{1}{(1+n\cos\theta)^2}+\frac{K_1}{c(1+n\cos\theta)^3}\right]d\theta = 0 \\ &\Rightarrow\frac{K_1}{c} = -\frac{\int_{\theta=0}^{\theta=2\pi}1/\left((1+n\cos\theta)^2\right)d\theta}{\int_{\theta=0}^{\theta=2\pi}1/\left(c(1+n\cos\theta)^3\right)d\theta}. \end{aligned} \tag{9}$$

On simplifying, we get a formula for calculating the first integration constant K_1:

$$K_1 = 2c\left(n^2-1\right)/\left(n^2+2\right). \tag{10}$$

Extreme oil pressure values as a function of the attitude angle θ are achieved if $dp/d\theta = 0$:

$$K_1 = -h = -c(1+n\cos\theta). \tag{11}$$

The first integration constant is related to the thickness of the oil film at the perimeter of the journal, where the maximum and minimum oil pressure is achieved:

$$h_m = (h)_{p=min} = (h)_{p=max} = -K_1 = \frac{2c(1-n^2)}{(n^2+2)}. \tag{12}$$

The attitude angle where the maximum and minimum pressure occur is given by

$$\cos\theta_m = -3n/\left(n^2+2\right). \tag{13}$$

The result of double integration is as follows:

$$p(\theta) = \frac{6\mu UR}{c^2}\frac{n(2+n\cos\theta)\sin\theta}{(n^2+2)(1+n\cos\theta)^2}+p_0 = \frac{6\mu UR}{c^2}\beta(\theta,n)+p_0. \tag{14}$$

The first integration constant was selected to meet the boundary condition $p_0(0) = p_0(2\pi)$ as is described by Dwivedy et al. The oil pressure distribution on the journal for $n = 0, 0.1, 0.2, \cdots, 0.9$ is shown in **Figure 5**. It should be noticed that the

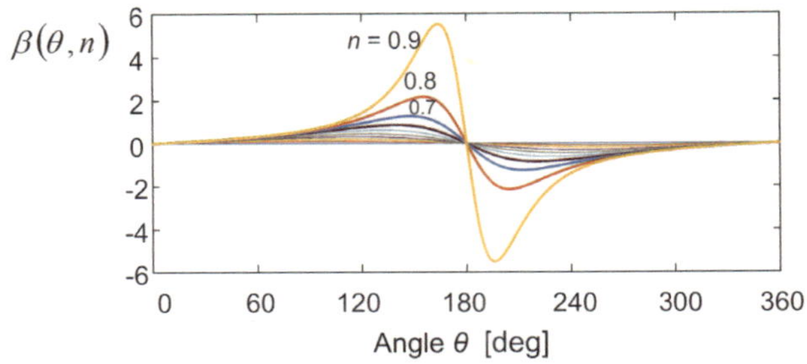

Figure 5.
Pressure distribution along the angular coordinate.

second integration constant has not any effect on the force excited by the oil pressure. The subatmospheric pressure creates a condition for the formation of the cavitation zones.

4.4 Fluid force

The forces acting on the journal in the centre of gravity along the bearing length of L can be calculated for the direction of the line of the centres and the perpendicular direction. Force in the direction of the line of centres is denoted as a direct force F_D, while force which is perpendicular to the line of centres is denoted as a quadrature force F_Q. Both these forces balance the gravity force G as is shown in **Figure 4**:

$$\begin{aligned} F_D &= \int_0^{2\pi} p_\theta \cos(\pi-\theta) LR \, d\theta = \\ &= F_N \int_0^{2\pi} \beta(\theta,n)\cos(\pi-\theta)\, d\theta = F_N \beta_D(n) \\ F_Q &= \int_0^{2\pi} p_\theta \sin(\pi-\theta) LR \, d\theta = \\ &= F_N \int_0^{2\pi} \beta(\theta,n)\sin(\pi-\theta)\, d\theta = F_N \beta_Q(n), \end{aligned} \tag{15}$$

where the force F_N can be designated as a nominal force because it corresponds to the maximum force according to the linear model ($n = 1$):

$$F_N = 6\mu U R^2 L / c^2. \tag{16}$$

Note that according to Eq. (14) the pressure on the part of the journal surface is negative, which is, in fact, a relative negative pressure. Since the pressure distribution is antisymmetric with respect to $\theta = \pi$, without evidence, it is clear that these formulas can be applied. Only quadrature force $F_Q > 0$ acts on the bearing journal, and the direct force are zero $F_D = 0$, as is shown on the left panel in **Figure 6**.

The nominal force that multiplies the dimensionless functions $\beta_D(n)$ and $\beta_Q(n)$ can be calculated with the use of the dimensionless Sommerfeld number S and the load P per unit of projected bearing area as follows:

$$F_N = 6\mu U R^2 L / c^2 = 6\pi S G, \tag{17}$$

where G is a gravity force.

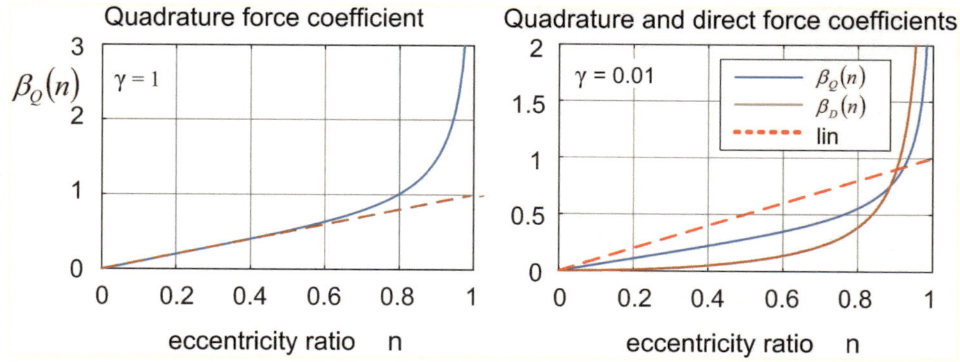

Figure 6.
Dependence of the direct and quadrature force on the eccentricity ratio.

For speeds ranging from 1.25k to 5k rpm, the force factor F_N varies from 0.46 to 1.8 kN. However, this force is reduced by multiplying the coefficients $\beta_D(n)$ and $\beta_Q(n)$, which depend on the eccentricity ratio n ranging from 0 to 0.33 (0.02/0.06). This case can only theoretically arise in an entirely flooded plain bearing with a vertical axis. The balance of forces F_D, F_Q, and F_Q allows to calculate an attitude angle α:

$$\alpha = \arctan\left(\beta_D(n)/\beta_Q(n)\right). \tag{18}$$

The presence of direct force can be explained, e.g., by the cavitation or the inability to achieve high vacuum, but the mathematical model is more complicated (Ferfecki) [8]. The lubricant flows through the bearing, but in the part of the bearing journal circumference where the pressure is below the barometric pressure, the lubricant can also be sucked. The magnitude of the negative pressure for $\pi < \theta < 2\pi$ is multiplied by a factor γ. Therefore the total force is given by the sum of integrals (Eq. (15)) as follows:

$$\begin{bmatrix} F_D \\ F_Q \end{bmatrix} = \int_0^{\pi} (\dots)\mathrm{d}\theta + \gamma \int_{\pi}^{2\pi} (\dots)\mathrm{d}\theta, \tag{19}$$

The effect of negative pressure reduction is demonstrated in the right panel of **Figure 6**. Negative pressure is limited to 1% of the magnitude of positive pressure for the angle interval of $0 < \theta < \pi$. The formulas for the calculation of the quadrature and direct forces contain the same factor $F_N = 6\mu U R^2 L/c^2$ and hence the dependence on the peripheral speed U and therefore on the rotor angular velocity. The coefficients $\beta_Q(n)$ and $\beta_D(n)$ differ considerably, in the experiment, the results of which are shown in **Figure 6**. The eccentricity ratio decreases approximately from 0.3 to 0.07 in the operation at the stable bearing position and stable lubrication. The diagrams confirm the linearity of the quadrature and direct force to eccentricity ratio up to 0.6. The $\beta_Q(n)$ and $\beta_D(n)$ coefficients can be approximated in this range as a linear function:

$$\begin{aligned} \beta_Q(n) &\approx qcn = qe \\ \beta_D(n) &\approx dcn = de, \end{aligned} \tag{20}$$

where q determines the quadrature stiffness $C_Q = 6\mu U R^2 L/c^2 \times q$ and d determines the direct stiffness $C_D = 6\mu U R^2 L/c^2 \times d$.

The stiffness in the directions of the Cartesian coordinates x, y and the attitude angle α which is defined in **Figure 5** can be obtained by substitution:

$$\begin{aligned} x(t) &= -e \sin\alpha \\ y(t) &= +\mathrm{e}\ \cos\alpha. \end{aligned} \tag{21}$$

The vector of the direct and quadrature forces depends on the coordinates x, y according to the following formula:

$$\begin{bmatrix} -C_D e \sin\alpha + C_Q e \cos\alpha \\ C_Q e \sin\alpha + C_D e \cos\alpha \end{bmatrix} = \begin{bmatrix} C_D x + C_Q y \\ -C_Q x + C_D y \end{bmatrix} = \begin{bmatrix} C_D & C_Q \\ -C_Q & C_D \end{bmatrix} \begin{bmatrix} x \\ y \end{bmatrix}. \tag{22}$$

The cross-coupled stiffness $D\lambda\Omega$ according to the Muszynska model corresponds to the expression $6\mu UR^2L/c^2 \times q$. The direct stiffness K is orderly less than the cross-coupled stiffness; however, the analytical calculation of the stiffness matrix shows the dependence on the rotational speed.

The damping matrix can be derived based on its relationship to the stiffness matrix according to the model that was designed by Muszynska:

$$\begin{bmatrix} D & 0 \\ 0 & D \end{bmatrix} \begin{bmatrix} \dot{x} \\ \dot{y} \end{bmatrix} = \begin{bmatrix} C_Q/\lambda\Omega & 0 \\ 0 & C_Q/\lambda\Omega \end{bmatrix} \begin{bmatrix} \dot{x} \\ \dot{y} \end{bmatrix}. \tag{23}$$

As is $\mathrm{D} = \mathrm{C_Q}/\lambda\Omega$, the damping coefficient D is a constant. The motion equation for the rigid rotor in the plain bearing is as follows:

$$\begin{bmatrix} M & 0 \\ 0 & M \end{bmatrix} \begin{bmatrix} \ddot{x} \\ \ddot{y} \end{bmatrix} + \begin{bmatrix} C_Q/\lambda\Omega & 0 \\ 0 & C_Q/\lambda\Omega \end{bmatrix} \begin{bmatrix} \dot{x} \\ \dot{y} \end{bmatrix} + \begin{bmatrix} C_D & C_Q \\ -C_Q & C_D \end{bmatrix} \begin{bmatrix} x \\ y \end{bmatrix} = \begin{bmatrix} F_X \\ F_Y \end{bmatrix}. \tag{24}$$

The sum of direct and quadrature forces must compensate for the gravitational force that does not depend on the speed of rotation. The suitability of this model is confirmed by Mendes [9].

If the attitude angle approaches the angle of $\pi/2$ in the radians, then the gravitational force is balanced only by the quadrature force. This state represents the stability limit. The experiment described in this article demonstrates the fact that hydrodynamic lubrication occurs when the eccentricity ratio n is decreased to below the value of 0.33, which happens after a certain speed margin has been exceeded. Unstable lubrication occurs during low revolutions when the eccentricity ratio is higher than the mentioned boundary. The problem of modelling the motion of the bearing journal at low rotational speeds raises the impossibility of determining the initial conditions.

5. Linear time-invariant mathematical model

The coordinate system in the complex plane for the bearing journal position is shown in **Figure 7**. A variable u is a control variable, and a variable r is a controlled variable. The controlled variable is a two-component coordinate of the bearing journal axis, while the control variable is a two-component coordinates of the bushing axis as is shown in **Figure 4**. Because both the variables indicate coordinates in the plane, then they can be considered as two-component vectors. The same meaning as the vector has a complex variables. The real part of this variable has the meaning of the x coordinate, while the imaginary part has the meaning of

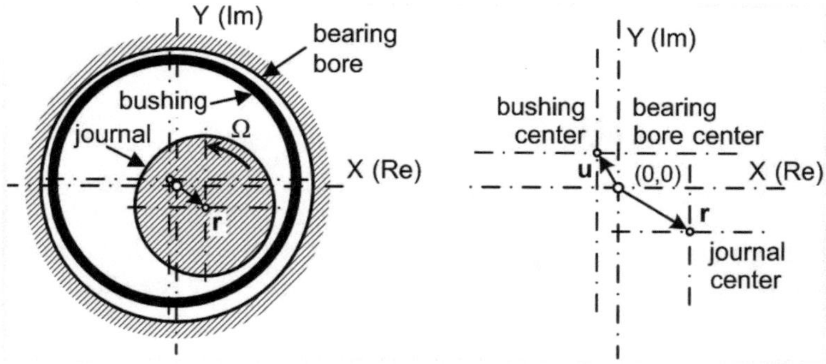

Figure 7.
Coordinate system in the complex plane.

the y coordinate, therefore it is possible to denote it as $r = x + jy$. The origin (0, 0) of the coordinate system in the complex plane is situated in the center of the mentioned cylindrical bearing bore.

There are many ways how to model journal bearings, but this paper prefers a lumped parameter model, which is based on the concept developed by Muszynska [10]. This concept assumes that the oil film acts as a combination of the spring and damper, which rotates at an angular speed $\lambda\Omega$ (variables will be defined hereinafter). The reason for using this concept was that it offers an effective way to understand the rotor instability problem and to create a model of a journal vibration active control by manipulating the bushing position with the use of actuators, which are a part of the closed-loop system composed of the proximity probes and the controller. If the bearing bushing is stationary ($u = 0$), then the equation of motion at the stationary rotation angular velocity is as follows:

$$M\ddot{r} + D\dot{r} + (K - jD\lambda\Omega)r = F_P \tag{25}$$

where M is the total rotor mass, Ω is the rotor angular velocity, K and D are specifying proportionality of stiffness and damping to the relative position of the journal center displacement vector, λ is a dimensionless parameter which is slightly less than 0.5, and F_P is an oscillating disturbing force defined by $\Delta F_P \exp(j(\omega t + \varphi_0))$, where ω is a synchronous or nonsynchronous excitation frequency and φ_0 is an initial phase. Excitation frequency is $\omega = 0$ for a static force and $\omega = \Omega$ for imbalance.

Force action of the oil film on the bearing journal can be modeled by Reynolds partial differential equations. The good accuracy of the Muszynska approximate model confirms Mendes and Cavalca [9]. Eq. (2) can be rewritten in matrix form

$$\begin{bmatrix} M & 0 \\ 0 & M \end{bmatrix}\begin{bmatrix} \ddot{x}(t) \\ \ddot{y}(t) \end{bmatrix} + \begin{bmatrix} D & 0 \\ 0 & D \end{bmatrix}\begin{bmatrix} \dot{x}(t) \\ \dot{y}(t) \end{bmatrix} + \begin{bmatrix} K & D\lambda\Omega \\ -D\lambda\Omega & K \end{bmatrix}\begin{bmatrix} x(t) \\ y(t) \end{bmatrix} = \begin{bmatrix} F_x(t) \\ F_y(t) \end{bmatrix}. \tag{26}$$

The entries of the stiffness and damping matrices according to Muszynska model and the calculation of these matrix entries using Reynolds equation agree except for very low rotor speed. Some entries are constants, and others are linear function of the rotational speed. Even entries of the damping matrix are similar. Coordinates of the bearing journal axis for the force of gravity $\boldsymbol{F_P} = 0 - jMg$ is given by the formula:

$$r_0 = Mg\frac{D\lambda\Omega - jK}{(D\lambda\Omega)^2 + K^2} \tag{27}$$

where g is gravitational acceleration. Stiffness of the oil film for static force increases proportionally with rotational speed of the bearing journal:

$$\frac{Mg}{r_0} = D\lambda\Omega + jK. \tag{28}$$

Movement of the bearing journal inside the bearing bushing may be unstable as is apparent from Eq. (26). This phenomenon is called a whirl. The threshold of stability in angular velocity of the rotor can be calculated by the Muszynska's formula:

$$\Omega_{CRIT} = \sqrt{K/M}/\lambda. \tag{29}$$

The equation of motion in the complex form for the rigid rotor operating in a small, localised region in the journal bearing with the movable bushing ($\boldsymbol{u} \neq 0$) is as follows (Eq. (26)):

$$M\ddot{\boldsymbol{r}} + D(\dot{\boldsymbol{r}} - \dot{\boldsymbol{u}}) + (K - jD\lambda\Omega)(\boldsymbol{r} - \boldsymbol{u}) = \boldsymbol{F}_P. \tag{30}$$

The Laplace transform specifies the transfer functions relating the displacement of the bushing to the displacement of the shaft $G_S(s)$, and the disturbance force to the displacement of the shaft $G_F(s)$ are given by

$$\boldsymbol{r}(s) = G_S(s)\boldsymbol{u}(s) + G_F(s)F_P(s), \tag{31}$$

where the mentioned transfer functions are as follows:

$$G_S(s) = \frac{\Delta\boldsymbol{r}(s)}{\Delta\boldsymbol{u}(s)} = \frac{Ds + (K - 1D\lambda\Omega)}{Ms^2 + Ds + (K - jD\lambda\Omega)} \tag{32}$$

$$G_F(s) = \frac{\Delta\boldsymbol{r}(s)}{\Delta\boldsymbol{F}_P(s)} = \frac{1}{Ms^2 + Ds + (K - jD\lambda\Omega)} \tag{33}$$

The active vibration control of journal bearings uses the bushing position as the control variable $\boldsymbol{u}$ and the shaft position as a controlled variable $\boldsymbol{r}$. The control variable is an output of a controller. The controller transforms an error signal computed as a difference of a reference (SP set point) and actual position of the journal. As is evident from the block diagram in **Figure 8**, the controller is of the proportional type with the gain of K_P.

Substituting $s = j\omega$ we can obtain frequency responses of the journal bearing system to harmonic oscillation of the bearing bushing position and disturbance force at the angular frequency $\omega = 2\pi f$ which can differ from the angular velocity Ω. Milling tool excites the strength of a frequency that is an integer multiple of the

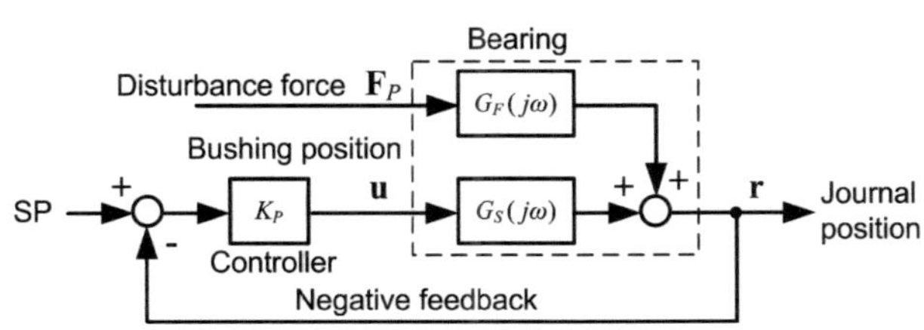

Figure 8.
Closed control loop.

frequency of rotation. The stead-state gain of $G_S(s)$ is equal to unit $G_S(j0) = 1$, while the stead-state gain of $G_F(s)$ depends on the rotor angular velocity as it results from the formula $G_F(j0) = 1/(K - jD\lambda\Omega)$. The reciprocal value of $G_F(j0)$ can be considered as the static stiffness of the oil film without influence of the active vibration control. Stiffness which is defined as a complex value determines the direction of journal axis displacement relative to the direction of the force. The radial force has theoretically identical direction with the journal displacement only at zero speed bearing journal.

The transfer functions relating the set point of the closed-loop system to the displacement of the shaft $G_w(s)$ and the disturbance force to the displacement of the shaft $G_D(s)$ are given by

$$G_w(s) = \frac{K_P G_S(s)}{1 + K_P G_S(s)} = \frac{K_P(Ds + (K - 1D\lambda\Omega))}{Ms^2 + (1 + K_P)Ds + (1 + K_P)(K - jD\lambda\Omega)}, \tag{34}$$

$$G_D(s) = \frac{G_F(s)}{1 + K_P G_S(s)} = \frac{1}{Ms^2 + (1 + K_P)Ds + (1 + K_P)(K - jD\lambda\Omega)}. \tag{35}$$

The stability margin can be calculated under assumption that the open-loop frequency transfer function $G_0(s) = K_P G_S(s)$ of the control loop in **Figure 8** is equal to -1. The frequency of the steady-state vibration at the stability margin is given by $\omega = \lambda\Omega$ and $K_P = \omega^2 M/K - 1$. If the feedback gain K_P is positive, then the maximal rotational speed Ω_{MAX} for the rotor stable behaviour is as follows:

$$\Omega_{MAX} = \Omega_{CRIT}\sqrt{K_P + 1}. \tag{36}$$

If the proportional controller is disconnected, i.e. $K_P = 0$, then the critical angular velocity Ω_{MAX} coincides with the critical frequency Ω_{CRIT} of the closed loop. Increasing of the stability margin for the rotational speed of the rotor is possible by introducing an additional feedback.

The reciprocal value of the transfer function (Eq. (35)) has the meaning of dynamic stiffness for radial force acting at the rotor:

$$C_D(s) = \frac{1}{G_D(s)} = Ms^2 + (1 + K_P)Ds + (1 + K_P)(K - jD\lambda\Omega). \tag{37}$$

If a static force is applied to a rotating journal, then the stiffness of the journal bearing is given by

$$C_D(j0) = \frac{1}{G_D(j0)} = (1 + K_P)(K - jD\lambda\Omega) \tag{38}$$

which means that the stiffness is $(1 + K_P)$ times greater than the journal stiffness without feedback.

Analysis of the effect of active vibration control on the stiffness of the bearing journal assumes a linear mathematical model. Practical calculation of matrix entries of stiffness **C** and damping **B** matrices

$$\mathbf{C} = \begin{bmatrix} C_{XX} & C_{XY} \\ C_{YX} & C_{YY} \end{bmatrix}, \mathbf{B} = \begin{bmatrix} B_{XX} & B_{XY} \\ B_{YX} & B_{YY} \end{bmatrix} \tag{39}$$

shows that the linear model does not differ substantially. The dependence of the matrix entries for the journal bearing of the test rig on the rotor rpm is given by the

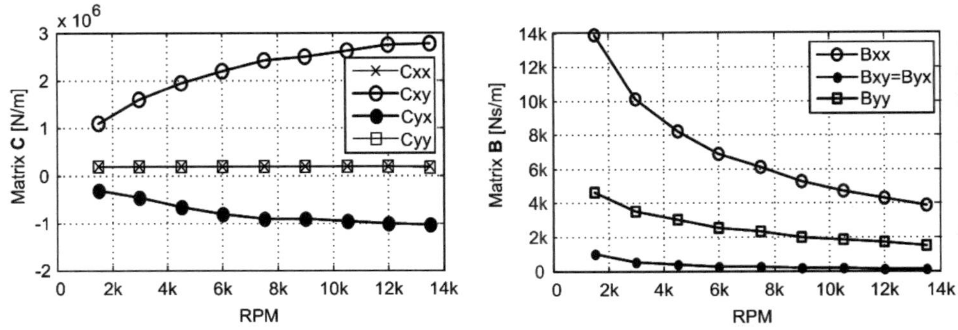

Figure 9.
Real stiffness and damping matrices according to the Reynolds model.

graphs in **Figure 9**. Pertinent stiffness and damping coefficients are obtained by solving Reynolds equation providing that the journal performs small harmonic motion in neighborhood of its equilibrium position.

6. Limits of the bearing bushing motion

The range of the manipulated variable, which is the position of bearing bushing and at the same time, the controller gain, determines the way to install piezoactuators. The equivalent circuit of the mechanical branch of the control loop for the horizontal direction is shown in **Figure 10**. Parameters that indicate stiffness in the scheme in **Figure 10** are associated to the individual elements of the control loop as follows: K_{PA} is for the piezoactuator, K_S is for the support, and K_{OR} is for the O-ring seal. The piezoactuator is a source of the mechanical travel u_x^* whose

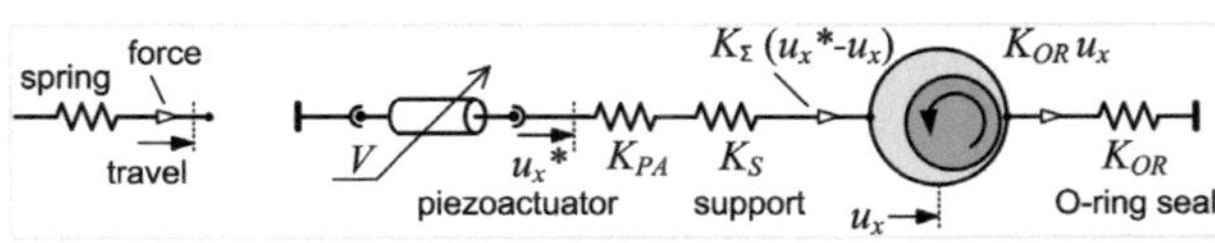

Figure 10.
Mechanical branch of the control loop.

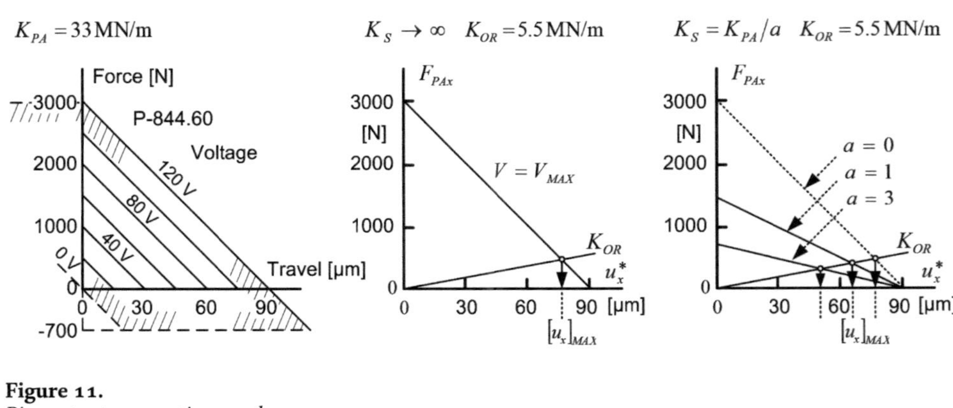

Figure 11.
Piezoactuator operation graphs.

magnitude depends on the voltage V. Displacement of the bearing bushing in the horizontal direction is designated as u_x.

The force produced by the piezoactuator balances the force effect of the oil film. Virtual motion of the unloaded piezoactuator is proportional to its control supply voltage $u_x^* = kV$. The resulting motion u_x of the bearing bushing also depends on the load force, as shown in the working graph of **Figure 11**.

Operation graph with all the limitations for the piezoactuator of the P-844.60 type is shown in the left panel of **Figure 11**. If the stiffness of sealing rings is taken into account and the stiffness of the support is assumed to be infinite, then the original range of motion of the bearing bushing is reduced to the size as follows:

$$[u_x]_{MAX} = \frac{[u_x^*]_{MAX}}{1 + K_{OR}/K_{PA}} \tag{40}$$

The effect of the stiffness of the O-ring seal is shown in the middle panel of **Figure 11**. Displacement of the bearing bushing is reduced from 90 to 77 μm for the given parameters of the control loop. The ultimate stiffness of the support affects the virtual stiffness of the piezoactuator as follows:

$$K_\Sigma = \frac{K_S K_{PA}}{K_S + K_{PA}} = \frac{K_{PA}}{1 + K_{PA}/K_S} = \frac{K_{PA}}{1 + a} \tag{41}$$

where a is a multiple. Maximum displacement of the bearing bushing respecting all influences is given by the following formula:

$$[u_x]_{MAX} = \frac{[u_x^*]_{MAX}}{1 + (1 + a)K_{OR}/K_{PA}}. \tag{42}$$

First experiments with an imperfect support showed displacement of the bearing bushing of about 20 microns at maximum electrical voltage to supply the piezoactuators. This happened at the beginning of the development when the support arrangement was provisionally extended due to the use of longer piezoactuators. The ideal solution is to install piezoactuators into the bearing housing.

7. Experiments with active control of journal bearings

Experiments with the active vibration control run for several years, while the hardware and software of the control system was upgraded step by step. We have improved design of the piezoactuator support, found suitable sensors for measuring the position of the bearing journal, upgraded the lubrication system, and improved the control algorithm. Properties of the active vibration system have previously been described in the paper [11], and now the main results will be described only, which relate to control the position of the bearing journal.

The instability onset of the bearing journal motion inside the bushing arises when crossing the threshold value of rotational speed Eq. (5). This phenomenon means that the steady-state rotation of the journal is not stable and the journal axis starts to whirl at the frequency, which is 0.42–0.48 multiple of the frequency of rotational speed of the rotor. Measurements in this article were carried out on the shaft with the radial clearance of 45 μm. Rotor speed increases according to a ramp function as it is shown in the left panel of **Figure 12**. The time history of the axis coordinates of the bearing journal is shown in the other panels of **Figure 12**. The x

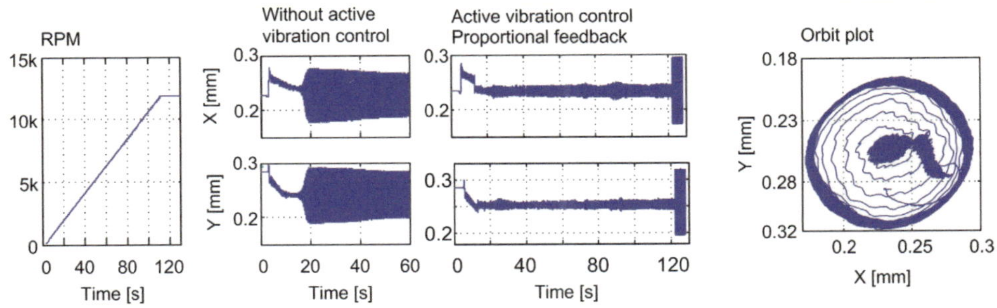

Figure 12.
rpm and the journal position as a function of time.

coordinate corresponds to horizontal direction, and the y coordinate is for vertical direction. Active vibration control is switched off for the time histories in the second panel. Instability occurs at about 2k rpm. This threshold of instability depends on the viscosity of the oil and the bearing clearance. Oscillations of the bearing journal position are limited by the journal clearance within the bushing.

If the active vibration control is switched on and rates of increase of rotational speed are identical for both measurements, the instability of the bearing occurs at the rotational speed about 12k rpm as is shown in the third and fourth panel of **Figure 12** from the left. Vibrations during the instable motion of the journal are also limited by the journal clearance within the inner gap of the bearing bushing.

The transient of the journal seems to be reverse for the vertical motion (y-axis) of the bearing journal in the second and third panel of **Figure 12**. The scale for the vertical motion is reverse in these figures, meaning upside-down. The relationship between horizontal and vertical motion of the journal shows the orbit of the journal axis in the rightmost fourth panel of **Figure 12**. The shape of the orbit is approximately circular when instability occurs.

Threshold of instability is increased six times now using a proportional feedback. According to Eq. (12) this multiple corresponds to the open-loop gain, which is equal to 35. Years ago, we achieved an increase in the threshold of instability only about by 70% for a piezoactuator support with insufficient stiffness. Such an increase of the instability threshold corresponds to the open-loop gain equal to 2.

The active vibration control is not turned on at 0 rpm of the rotor but after finishing a transient process, which ends by lifting the journal to approximately the middle position in the vertical direction which takes approximately 15 seconds for the given rate of the increase of speed.

Through the experiment under specific conditions, the observed onset of instability was at 8450 rpm for control only in the x-axis direction and at 7100 rpm for control only in the y-axis direction [12]. It confirms the rule that static load delays the onset of instability at higher speeds. Control in both directions is required if the direction of the radial force may change or if the rotor has a vertical axis, i.e., the radial force is missing.

The linear proportional controller was used for active vibration control for measurements presented in **Figure 13**. Parametric excitation means that at least one parameter of the system varies periodically in time according to a sinusoidal function, as was suggested by Tondl and Dohnal [13, 14]. The gain of the proportional controller was selected as this varying parameter. The system becomes nonlinear and nonstationary. The gain of the proportional controller is given as follows:

$$K_P = K_{P0}(1 + \alpha \sin(\omega_0 t)), \qquad (43)$$

where α is dimensionless amplitude of excitation, K_{P0} is static gain factor, and ω_0 is angular frequency of excitation.

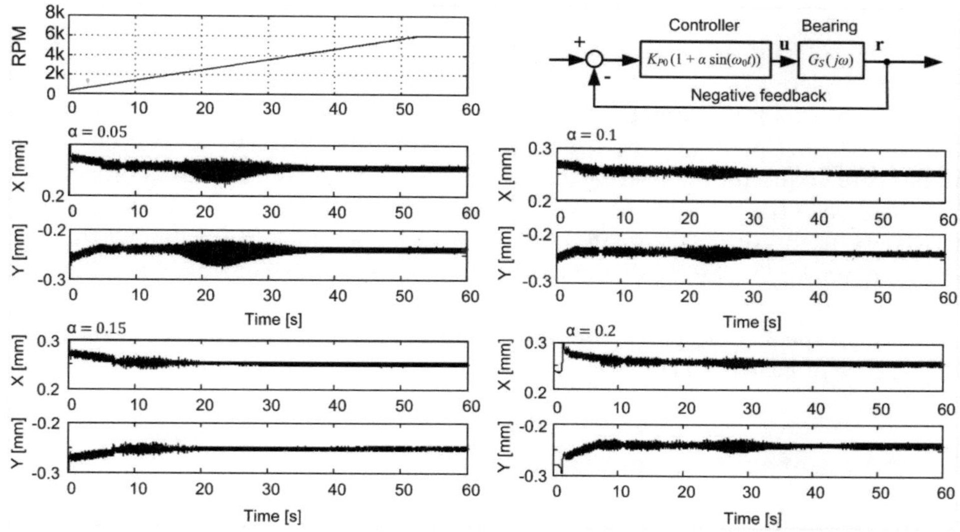

Figure 13.
The journal position as a function of time for tests with active vibration control on.

Dohnal [14] has solved a similar problem for magnetic bearings. Our experiments on the test bench were conducted for the following amplitudes of excitation α = 0, 0.1, 0.15, and 0.2. The static gain was the same as the gain of the previous experiments with the linear controller. The excitation frequency was selected 30 Hz, which is approximately equal to the frequency of vibration at the low rpm. Rotor speed increases according to a ramp function as is shown in the first left panel of **Figure 13**. The effect of the amplitude of the parametric excitation on the journal movement during rotational velocity run-up is shown in other panels of **Figure 13** [11]. The best choice of the excitation amplitude is α = 0.15, which is the position of the journal almost without oscillations. The amplitude of the residual oscillation of the journal does not exceed 8 μm. Precision ball bearings (so-called deep groove ball bearings) which are offered by SKF have a radial clearance (radial internal clearance C2) to a diameter of 30 mm in the range from 1 to 11 micrometers. The maximum rotational speed of the 206-SFFC bearing type is only 7.5k to 13k rpm.

8. Reducing mechanical power losses in actively controlled bearings

Power losses in the journal bearings were estimated from the electric power which is consumed by frequency convertor and motor. Dependence of electrical power upon rotational speed of the motor was measured with and without active control as it is shown in **Figure 14**. Basic power consumption of the motor and

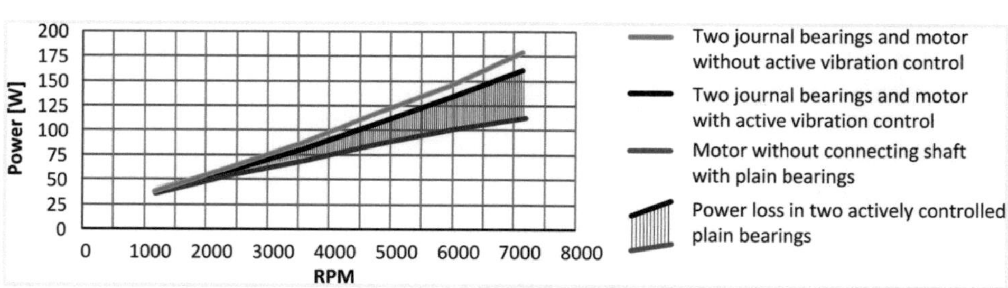

Figure 14.
The electric power consumed by the frequency converter, motor, and bearings.

frequency convertor was measured with the disconnected clutch between the motor and rotor; it means that the bearings were inoperative. The friction loss of a pair of bearings at 7k rpm is 66 W in an unstable operation, and if the active vibration control is on, then the friction loss is of only 48 W. The active vibration control reduces the friction losses of journal bearings by 27%. The bearing clearance amounts to 90 μm for the bearing journal of the diameter 30 mm. As a lubricant the hydraulic oil of the OL-P03 type (VG 10 grade, kinematic viscosity 2.5 to 4 mm^2/s at 40°C) was used. All tests were undertaken at ambient temperature about 20°C. For small power loss by friction in the bearings, the actively controlled bearings can be used in systems for storing the kinetic energy as they are flywheels that spin at high speed. Longer life compared with roller bearings is another advantage of this type of bearings [11].

9. Stiffness of actively controlled bearings

The bearing bushing is suspended on a pair of piezoactuators, and the bearing journal is supported by an oil wedge. According to catalogue data, we used a linear piezoactuator, which is able to generate force of 3 kN in pressure or 700 N in tension on the track in the range of 90 μm. These parameters correspond to the piezoactuator stiffness of 33 MN/m. Stiffness of the force transducer is 2000 MN/ m, which is two orders of magnitude higher than the stiffness of piezoactuators. Stiffness of the O-ring seal is 5.5 MN/m which increases the stiffness of the piezoactuator by this value. Force is transmitted to the bearing journal through the oil film. Based on the simulations, it can be estimated that the direct stiffness (C_{XX}) of the oil film in neighborhood of the central position within the bushing bore is of 185 kN/m and the quadrature stiffness (C_{XY} and C_{YX}) is still an order of magnitude larger, and increases proportionally with the rotational speed; see Eq. (39) and **Figure 9**. This stiffness increases by many orders of magnitude if the journal is approaching the bearing bushing wall. Stiffness of the journal support is defined first of all by stiffness of the oil film. A steady-state error in a noncontrolled bearing originates in a radial load which can be considered as a disturbance. A proportional controller which governs a system of journal bearings in the closed-loop with an open-loop gain K_P reduces the steady-state error $(1 + K_P)$ times which results in increase of the oil wedge stiffness $(1 + K_P)$ times compared to the design without a feedback. An integration controller reduces steady-state error to zero, which corresponds to a theoretically infinite stiffness. The use of a derivative component of the controller was also analysed, but additive noise would reduce its effect [15]. Allowable forces, however, are limited by the load capacity of the piezoactuators. The pressure force is thus less than 3 kN. Notice that on the market there are piezoactuators enabling to generate forces up to 20 kN.

Stiffness of precision rolling bearings ranges from 100 to 200 kN/m, regardless of the load, while the stiffness of hydrodynamic bearings in neighborhood of central position (low load) is of the order of several kN/m. However, with active control, the stiffness can increase as much as 35 times, i.e., it can achieve values around 100 kN/m, which is comparable to that of precision ball bearing.

10. Conclusion

Experiments prove the correctness of the theoretical prediction which refers to the extending of the operating range of plain bearings when active vibration control is used. The performance of the actively controlled bearing was tested on the test rig. The bearing diameter is 30 mm, and the length-to-diameter ratio is equal to about 0.77. The radial clearance of the journal is 45 μm and the very low viscosity oil is used. This combination causes instability of the oil whirl type from the rotational speed of 2k rpm. The active vibration control extends stable operating rotational speed up to 12k rpm, i.e., six times. Also the stiffness of the bearing journal increases significantly during a displacement from equilibrium position. The friction loss of a pair of bearings at 7k rpm is 66 W in an unstable operation, and if the active vibration control is switched ON, then the friction loss is of only 48 W. The active vibration control reduces the friction losses by 27%. The linear proportional controller was used for the active vibration control. The quality of control has been enhanced with the use of periodic changes of the controller gain, which is known as a parametric excitation. The effect of this way of control reduces the journal residual oscillation to the limits which does not exceed 8 μm. This amplitude is comparable with the radial clearance of the ball bearings of the deep groove type. The experiments with the time-periodic changes of the controller gain confirm the positive effect on the vibration response.

Acknowledgements

This work was supported by the European Regional Development Fund in the Research Centre of Advanced Mechatronic Systems project, CZ.02.1.01/0.0/0.0/16_019/0000867 within the Operational Programme Research, Development and Education and the project SP2020/57 Research and Development of Advanced Methods in the Area of Machines and Process Control supported by the Ministry of Education, Youth and Sports. This publication was issued thanks to supporting within the operational programme research and innovation for the project: "New generation of freight railway wagons" (project code in ITMS2014+: 313010P922) co-financed from the resources of the European Regional Development Fund.

Author details

Jiří Tůma[1*], Jiří Šimek[2], Miroslav Mahdal[1], Jaromír Škuta[1], Renata Wagnerová[1] and Stanislav Žiaran[3]

1 Department of Control Systems and Instrumentation, VSB Technical University of Ostrava, Ostrava, Czech Republic

2 TECHLAB Ltd., Prague, Praha, Czech Republic

3 Mechanical Engineering Faculty, The Slovak University of Technology in Bratislava, Slovakia

*Address all correspondence to: jiri.tuma@vsb.cz

References

[1] Budynas RG, Nisbett JK. Shigley's Mechanical Engineering Design. 9th ed. New York: McGraw-Hill; 2011

[2] McKee SA, McKee TR. Journal bearing friction in the region of thin film lubrication. SAE Journal. 1932;**31**: 371-377

[3] Simek J, Tuma J, Skuta J, Klecka R. Unorthodox behavior of a rigid rotor supported in sliding bearings. In: Proceedings of the Colloquium Dynamics of Machines. Prague: Institute of Thermomechanics; 2010. pp. 85-90

[4] Simek J, Tuma J, Skuta J, Mahdal M. Test stand for affecting rotor behavior by active control of sliding bearings. In: Proceedings of the Colloquium Dynamics of Machines. Prague: Institute of Thermomechanics; 2014. pp. 151-156

[5] Tůma J, Šimek J, Škuta J, Los J. Active vibrations control of journal bearings with the use of piezoactuators. Mechanical Systems and Signal Processing. 2013;**36**:618-629

[6] Wagnerová R, Tůma J. Use of complex signals in modeling of journal bearings. In: 8th Vienna International Conference on Mathematical Modeling. Vienna; Austria: MATHMOD; 2015. pp. 520-525

[7] Dwivedy SK, Tiwari R. Dynamics of Machinery, A Lecture Notes. Guwahati, India: All India Council of Technical Education; 2006

[8] Ferfecki P, Zapomel J. Investigation of vibration mitigation of flexible support rigid rotors equipped with controlled elements. In: 5th International Conference on Modeling of Mechanical and Mechatronic Systems. Vol. 48. Zemplinska Sirava; Slovakia: MMaMS; 2012. pp. 135-142

[9] Mendes RU, Cavalca KJ. On the instability threshold of journal bearing supported rotors. International Journal of Rotating Machinery. 2014;**2014**: 351261

[10] Muszynska A. Whirl and whip – rotor/bearing stability problems. Journal of Sound and Vibration. 1986;**110**(3): 443-462

[11] Tůma J, Šimek J, Mahdal M, Škuta J, Wagnerová R. Actively controlled journal bearings. In: Ecker H, editor. Proceedings of the 12th SIRM Conference. Graz, Österreich; 2017. pp. 443-462

[12] Los J. Mechatronické systémy s piezoaktuátory. VSB Technical University of Ostrava. PhD. Thesis.; 2016

[13] Tondl A. Quenching of Self-Excited Vibrations. Prague: Academia; 1991

[14] Dohnal F, Markert R, Hilsdorf T. Enhancement of external damping of a flexible rotor in active magnetic bearings by time-periodic stiffness. In: Proceedings of the SIRM, Internationale Tagung Schwingungen in rotierenden Maschinen. Darmstadt, Deutschland; 2011

[15] Víteček A, Tůma J, Vítečková M. Stability of rigid rotor in journal bearing. In: Transactions of the VŠB, Mechanical Series. No. 2. Vol. LIV. Technical University of Ostrava; 2008. pp. 159-164

Section 3

Application Usage

Chapter 7

Application of Machine Tools in Orthoses Manufacture

Karlo Obrovac, Miho Klaić, Tomislav Staroveški, Toma Udiljak and Jadranka Vuković Obrovac

Abstract

CNC technology is widely used in the manufacture of medical products. An area in which CNC technology has proven to be extremely useful and innovative is Orthotics and Prosthetics (O&P). O&P laboratories are engaged in the manufacture of individual orthoses and prostheses. The usual manual manufacture of such products takes a long time and requires tremendous experience and skill. In this regard, any engineering solution that improves the quality of the production process; reduces production time, production costs, and physical human labor; and at the same time improves the environmental conditions of the production environment will be desirable. Various designs of CNC machine tools for the manufacture of orthoses or molds for their production are in use today. In most cases, customized commercially available numerical control lathes and milling machines are used, as well as industrial robotic arms, but there are also highly specialized designs. For the mentioned purpose, we also encounter the application of additive manufacturing (AM) devices. Due to the fact that issuing of orthoses is often the subject of cost reduction in healthcare systems, the pursuit of production systems that will be cost-effective and functional, easily implemented, and used primarily in small manufacturing practices is imperative.

Keywords: CNC, machine tools, orthoses, O&P

1. Introduction

O&P is a medical-technical profession involved in the manufacture of orthotic devices, prostheses, and similar products. Orthoses are aids that serve to compensate for a functional deficit, while prostheses serve to replace a missing part of the body.

The need for such aids is as old as the human species. Due to trauma, illness, or other circumstances, a person is exposed to the possibility of developing a disability, loss of working ability, and, generally, a reduction in the quality of life. With the application of various medical treatments, people have sought to solve this problem by using available things from their environment, or in this sense made simple props and devices, with which they have compensated the negative effects of their present deficit. Numerous events in the past have encouraged greater demand and need for a more comprehensive approach for the design and application of such aids, thereby improving their performance and quality.

IntechOpen

It is well known that consequences of traumas, deformities, old age, certain pathological processes, or impaired human biomechanics can produce a deficit of the locomotor system. In these conditions, orthoses are used as the main aids in daily life.

Orthoses are roughly divided into:

- Craniofacial orthoses
- Trunk orthoses
- Upper extremity orthoses
- Lower extremity orthoses
- Orthopedic inserts (**Figure 1**)
- Orthopedic shoes

The application of modern technologies has enabled the simplification of otherwise demanding procedures such as taking measurements and making the orthosis itself.

In the standard procedure, these issues depend entirely on the level of expertise and experience of the individual or the laboratory in which the production takes place.

Mostly it is about a manual technique of manufacturing using available tools and machines, which are used to modify the plaster mold.

If we take spinal orthosis manufacturing as an example, this mold has the shape of a human torso (**Figure 2**). It is made by pouring gypsum into the primary plaster mold, which is formed by solidifying the plaster bandages wrapped around the body, which is a particularly unpleasant procedure for the patient [1].

The mold obtained in this way weighs tens of kilograms, is difficult to manipulate, fragile, and, what is particularly important, rigid; so, it is very complicated to carry out the desired modifications of the surface. Due to the imperfections of such a procedure, inadequate products are often made.

A team of experts in the fields of medicine, mechanical engineering, computer science, and other related professions participates in the orthoses design and its production for the patient. The application of CAD/CAM technologies has significantly simplified this process for the past 30 years.

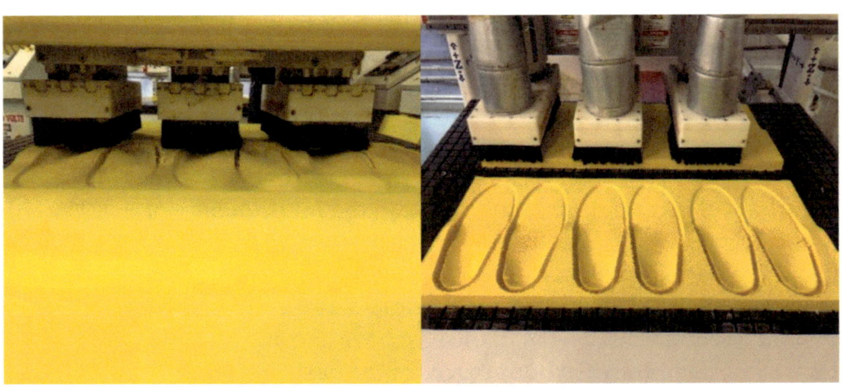

Figure 1.
CNC milling machine with three spindles.

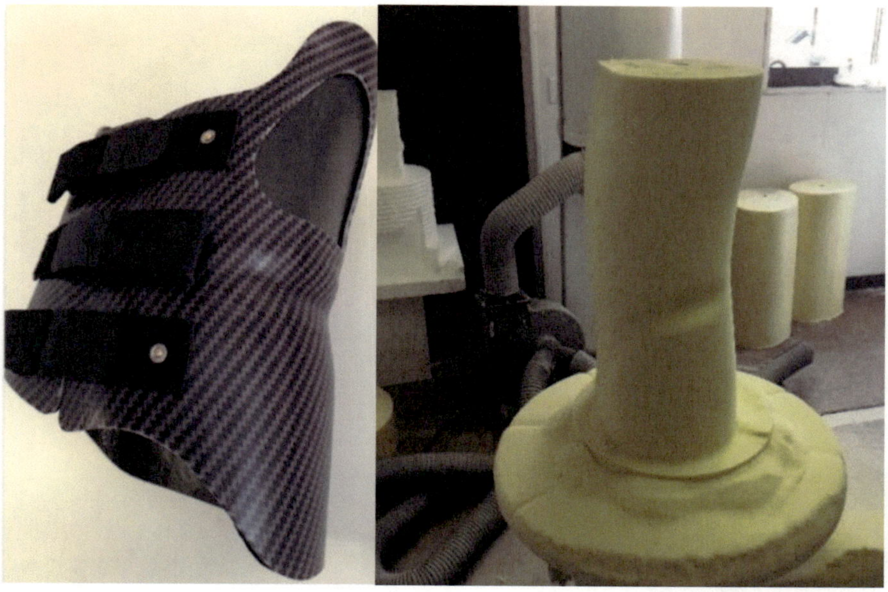

Figure 2.
Spinal orthosis and mold.

Although it is used in the manufacture of all types of orthoses, it is certainly most used for the purpose of making orthopedic insoles with the typical use of three-axis CNC milling machines (**Figure 3**).

Orthopedic insoles are made by direct machining of the materials (usually EVA—ethylene vinyl acetate) or by making molds from materials that include expanded polyurethane, expanded polystyrene, or MDF (**Figure 4**). The insoles are characterized by smaller dimensions (especially thicknesses) while the same cannot be claimed for other orthoses.

In the process of orthopedic insole design, the following is commonly considered:

- Clinical information
- Data obtained from diagnostic procedures (MRI, CT, standard radiograms, EMG, ENG, force, and pressure measurements) (**Figure 5**)

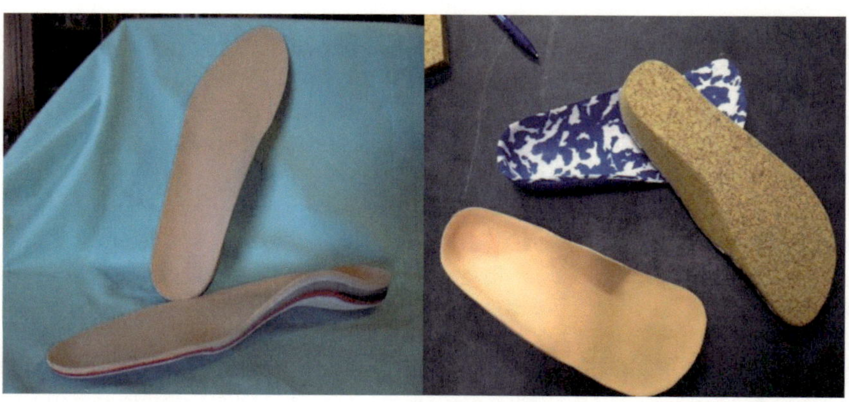

Figure 3.
Orthopedic insoles.

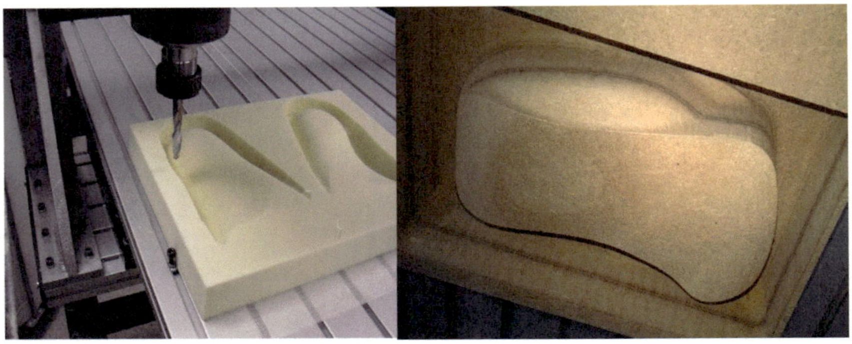

Figure 4.
Molds for foot orthoses production.

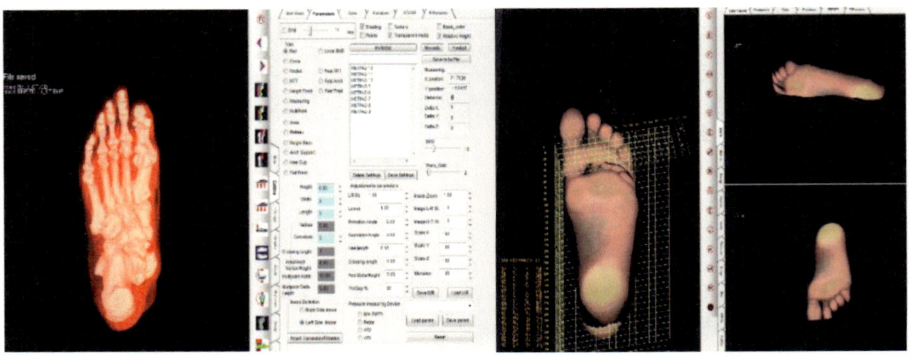

Figure 5.
Design of foot orthoses using CAD software.

- Biometric data, including measurement of the range of motion of an individual body part
- Geometric description of the topography of a body part of interest obtained by the 3D scanning
- Thermographic images
- Other available information (body posture and gait analysis, stabilometry, etc.)

While in the standard procedure, these data (e.g., standard radiograms) are reviewed and decisions are made on modifications of plaster molds based on them, in the digital procedure they are imported in digital form into the appropriate CAD program as data records, where they are used as a basis for modification or navigation during design (**Figure 6**).

Upon completion of the design, tool paths for the selected manufacturing process and hardware are generated, and then either an orthosis or a mold for orthosis is made (**Figure 7**). Regarding the application of digital solutions in the manufacture of molds for orthoses, additive technologies and machining technologies are used.

The application of additive technologies [2] in the manufacture of orthoses is of recent date (**Figure 8**). This approach far surpasses other technologies in the ability to create complex shapes, but despite great technological advancements both in materials and hardware solutions, as well as initially promising results, it still

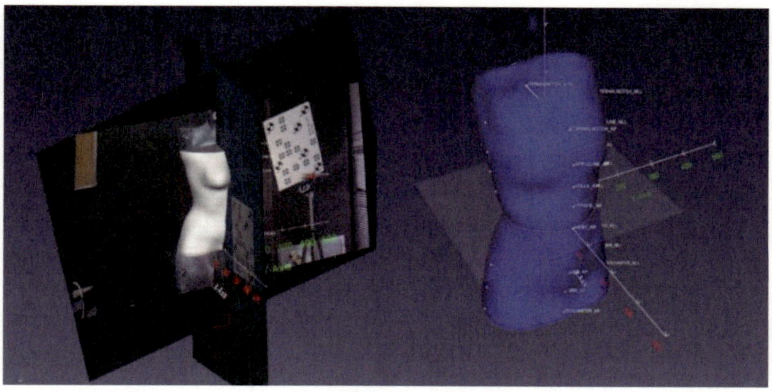

Figure 6.
Spinal orthosis design—CAD.

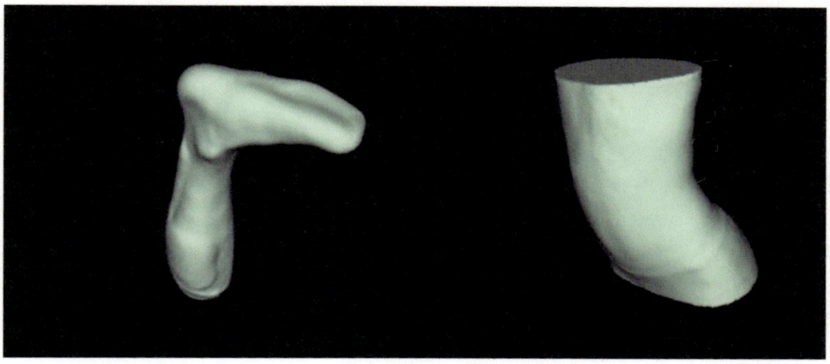

Figure 7.
Completed design for AFO and spinal orthosis.

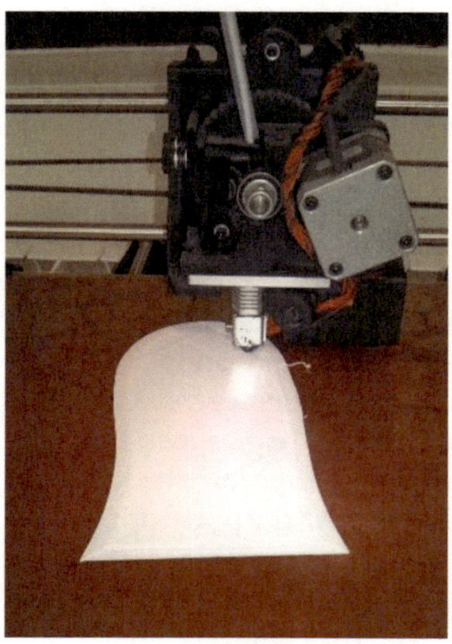

Figure 8.
Additive manufacturing of foot orthosis.

requires more significant research efforts and improvements to become a full-scale manufacturing technology. It is expected that new solutions will overcome the present problems of mechanical properties of the product, the cost of production of larger volume forms, as well as the time of making orthosis.

These issues are raised particularly when it comes to implementing such a manufacturing system in smaller orthotic laboratories engaged in the manufacture of various types of orthotic devices, and where the implementation of new technological solutions is of great significance for business stability.

However, even after more than 30 years of application of digital technologies, the manufacturing procedure itself remains a significant problem—it is still expensive, not practical enough, often too complicated, and it is common to question the feasibility of investing in its implementation.

With CAD/CAM technology, the biggest financial share is the price of the machine. There are a number of special solutions currently present on the market with the primary purpose of making molds for orthoses. The use of such solutions certainly simplifies the manufacturing process, but they are expensive and are usually found in larger laboratories.

The performance of such machines is often related to the production of only certain types of orthoses, and they need to be modified for other applications because they do not have the proper functionality, which is neither easy nor affordable.

In addition, the kinematics of such machines is often inappropriate for machining molds of more complex geometry, which requires additional human labor.

2. CAD/CAM

The introduction of CAD/CAM systems into the production of orthoses has led to a great advancement in orthotic practice [3].

Its application has brought a major change over the classical (traditional) approach, in which orthoses are made manually.

The use of digital technology in O&P practice is mainly based on results from a volumetric measurement. The classic volumetric measurement necessary for the design of orthopedic inserts is obtained by taking foot impressions in polyurethane foam and plaster molding. In both cases, it is implied that plaster needs to be poured into resulting impressions and one needs to wait for it to solidify. Issues related to this process include the fact that the foam box is brittle, that taking the impression requires some experience (otherwise, a distorted foot shape is obtained), and that the plaster molding is only as good as it was carefully and qualitatively shaped over the foot surface (because it is taken in the non-weight-bearing position). Finally, the molds obtained are rigid and require considerable manual work to achieve proper shape, over which the material can be thermoformed to obtain an insole shell. Sometimes, it is not possible to completely modify the mold while retaining its baseform, because it is rigid, and it is often necessary to add material to one part of the mold and to subtract it in the other part. Also, it is complicated to add and to form corrective elements (e.g., a metatarsal pad feature), because during the molding the shape changes and thus loses its initial form; so, such a "design" depends largely on the experience of the practitioner. With this process, it can take up to 6 h to obtain a mold that is suitable for thermoforming the material. An additional problem is the transportation of the obtained plaster moldings or foam boxes since they are brittle, and the transport itself requires a considerable amount of time and human labor. In contrast, digital technology enables rapid 3D scanning of the entire foot and ankle surface. The information provided is in digital form and can be easily transmitted to a remote location. In such a digital form, the data are fed into a CAD program where they can be manipulated and modified in a much faster and easier way, and without the expense of materials. Such a process, which involves acquisition and design, takes an average of up to 20–30 min depending on the experience of the designer. The use of digital technologies generally provides for the possibility that the locations where scanning, design, and production are performed are different. In practice, we often find the cases where data acquisition and design are performed on one location while orthoses manufacturing is made in a central laboratory that is technically equipped to produce orthoses on a larger scale.

This approach certainly has its advantages and disadvantages. The advantages are found in the fact that the service seeker does not need to possess production equipment. However, due to the large volume of production in such laboratories, there is a potential risk of extended insole delivery time, variations in the quality of the product, and other disadvantages of economic nature.

Modern technologies also allow the use of other information in the design of orthoses, which is especially important in the complicated cases that are often encountered in clinical practice. By using them, it is possible to obtain the surface that is a basis for the design, to conduct simulations of the influence of the insole surface configuration on the displacements of anatomic structures (using models obtained by reconstruction from existing sets of DICOM images obtained at CT and MRI diagnostic procedures; e.g., to define the surface of the skin border and surrounding environment; it is also possible to reconstruct the internal bone structures and use them for navigation during design), and to carry out FEA analysis (e.g., using the abovementioned reconstructed anatomical models and information obtained from the pedobarographic measurements). The quality of design can also be improved by the use of other information such as thermographic images, as well as those that give us additional insight into the shape or function of the part of the body of interest to which the orthosis will be applied.

The application of the CAD/CAM system, unlike the classical methodology, allows both the manufacture of molds (which can be recyclable material and are lighter than plaster) and manufacture of orthoses by direct material machining. This fact is especially emphasized in the production of an orthopedic insole base. The use of CAD/CAM technology also makes it possible to machine both the top and the bottom sides of the insole, either by using special machines that allow for simultaneously machining both insole sides, or in two setups, each insole side in separate setup. This makes it possible to realize an insole in a high degree of completion and with a minimal need for additional human labor. **Table 1** shows the comparison of the methods of making orthopedic inserts using classical and digital procedures. The data are based on years of experience of the authors.

Procedure	Classical	CAD/CAM
Acquisition of information about the shape of the part of the body to which the orthoses will be applied (min–max time)	15–60 min	5 s–8 min
Design and manufacture of the orthoses mold (min–max time)	1–6 h	5–30 min
Production of the orthopedic insoles by molding and finalization (min–max time)	20–40 min	20–40 min
Ability to create the orthoses without mold	Very low (+)	Very high (+++++)
Creating the orthoses basis by direct material machining (min–max time)	2–3 h	5–15 min
Repeatability of the appearance and properties of the orthoses	Low (++)	Very high (+++++)
Cost of manufacture (excluding the cost of machines and their depreciation)—material, time, design, supplies (average)	100–250 €	50–130 €
Ability to achieve the targeted mechanical properties of the orthoses	Low—high (+++)	Very high (+++++)

Table 1.
Comparison of features of classic and CAD/CAM approaches in the production of foot orthoses.

The data presented undoubtedly point to the advantages that modern digital technologies provide in acquiring the information needed to define orthoses and of applying modern manufacturing technologies over the classical process. Digital volumetric measurement takes shorter time, and since it is most commonly performed with non-contact scanners, it does not require the consumption of materials, and given the digital nature of the information obtained, there is a minimal possibility of damaging them through transmission. The CAD design is carried out in the environment where many editing tools are provided. Using these tools, targeted corrections are achieved, and if not satisfied, this action may be terminated and design returned to the previous state. CAD module enables measurements (e.g., angles, Euclidean and geodesic distances, ranges, curvatures, area, etc.) that could be used to predict the volume and shape of the workpiece material and other materials needed to assemble the orthoses (e.g., appearance and surface of the insert top cover material that will be applied to the insole top aspect, or the size and contour of the material that will be molded over the mold by thermoforming). This material can also be cut using a CNC machine with a tangential knife, laser head, or a machining tool. The given approach saves material and prevents possible errors. By using appropriate software tools, it is also possible to smooth the surface in such a way that its main features will be preserved, while this is not the case in the classical procedure. Here, the mold is mostly obtained by pouring plaster into the impression of a foot in foam. After a considerable amount of time for the mold to solidify, there follows a physically demanding procedure for shaping the mold to achieve an adequate form with adjustments and a smooth surface. This involves the addition and subtraction of material on the rigid mold, which inevitably loses its shape. Measurements on such a mold are more complicated and more inaccurate. As previously mentioned, applying CAD/CAM in insole design may also include machining of the insole bottom, which results in an insole ready to use with just minimal additional processing. In the classical approach, forming the insole bottom is very rarely applied because it is extremely time-consuming. One very important feature of the CAD/CAM technology in making an orthosis is the ability to make a multiple copies of the orthosis mold or the orthosis itself. Using CAD/CAM technology, repeatability is very high, and all copies of the orthosis will have exact shape. This is, again, not the case in the classical procedure. The procedure here depends on the experience of the expert who is making the orthosis, the quality of the mold previously used, and other related issues. For this reason, users of the classically made orthosis copy often suggest that there is a significant difference in the properties and appearance of the orthosis when compared with the previously made one, and additional modifications are often required. Creating an orthosis using CAD/CAM technology also makes for great savings in material, time, and energy. One of the biggest reasons for this is that during the design of the orthosis, several parameters can be considered, based on which it is possible to select the appropriate blank size (whether for direct orthosis machining or for orthosis mold), to adjust the machining parameters, and to optimize the trajectories to obtain appropriate characteristics and quality of the surface. Also, the need for human labor has been sharply reduced, so the cost of manufacturing is almost halved compared to making an orthosis with the classical procedure. A particular aspect of the application of CAD/CAM technology is the technical quality of the product and the question of achieving the desired mechanical properties. If one considers the classical process of insole making, it is mostly carried out by thermoforming the material on the mold in a vacuum press. The heated (usually polymeric) material is formed over the mold surface. In this process, it changes its nominal properties and often loses those that are needed to achieve an appropriate therapeutic effect. This is especially manifested in conditions that are very common in clinical practice, which require the use of very soft materials (e.g., diabetes, rheumatoid arthritis, etc.). Since these materials are the most often porous, the pressing process in a vacuum press pushes the air out of the cells of material, and they becomes tiffer. Direct machining of materials in the CAD/CAM process minimizes loss of their properties, especially if the parameters are well adjusted.

A particular issue is the achievement of certain mechanical characteristics of the orthosis by determining the thickness of the material in a particular region. This is a common requirement and is much easier to implement with CAD design than through manual forming in the classical procedure.

From mentioned above and other indicators, it can be concluded that the implementation of the CAD/CAM procedure has contributed that the process of manufacturing an orthosis became:

- More massive
- More practical
- More specific
- Faster
- More predictable
- More accurate
- More repeatable and
- The application of the appropriate CAD program enables to carry out such corrective actions that cannot be easily achieved in the classical approach.

Generally, the benefits resulting from the application of digital technologies in manufacturing orthoses include:

- Portability
- Reduced manufacturing complexity
- More time available to work with the patient
- Reduction of storage and production space
- Simplicity of copying and manufacturing the same orthoses
- Simplicity of upgrading the production process
- Possibility of distributed production
- Reduced labor costs
- Higher productivity

The particular benefit of using digital technologies is the achievement of more humane working conditions, that is, excluding people from jobs that are:

- Physically demanding
- Boring
- Repetitive
- Dangerous

The introduction of the CAD/CAM technology into O&P has led to faster and higher manufacturing of quality orthoses and has accelerated and simplified the procedure itself for both the patient and the orthotic practitioner [1].

3. Literature review

Although there are numerous descriptions of manufacturing orthoses in the literature, a review of their production using CNC machine tools is relatively scarce. We can find a large number of articles that generally describe the application of the CAD/CAM technology in orthotics and prosthetics, but there is little information with a more detailed description of the technical design of such systems.

- Orthotic practice has been radically changed by the implementation of digital technologies [3].
- Particular progress in orthotic practice has been achieved through the implementation of contactless scanners of human body parts and CAD/CAM technology. The application of scanners reduces the time of defining the shape of a body part. Also, it eliminates physical work involved in the process of acquisition of the human body geometry, enables storing of information in digital form without the accompanying problems of mold storage, enables transmission of data over the network, and provides high accuracy of acquired results [1, 3–5].
- Although the application of the CAD/CAM technology is unlikely to completely eliminate the need for human labor, its application in the production of orthoses has made a major technological breakthrough, and progress in technical quality and efficiency of the orthosis performance, with overall simplification of the procedure, and reduction of the need for otherwise hard work [5, 6].

It is evident that larger workpiece dimensions require more space, larger machines, and therefore more financial appropriations.

In addition to the three-axis, we also often encounter four-axis milling machines, turning centers and multitasking machine tools. Industrial robots are also increasingly used in O&P practice [7–11] (**Figure 9**).

However, the mentioned solutions still do not fully cover the main requirements of the practice: availability, greater efficiency, flexibility, and applicability within healthcare institutions.

The application of digital production systems is closely related to the beginning of the 3D scanning application. The use of a 3D scanner allows the reduction of time for defining the shape of a body part of interest (as well as elimination of the physical labor used to define the shape of an orthosis or its mold in the classical procedure) and load the results of scanning into a CAD program (**Figure 10**).

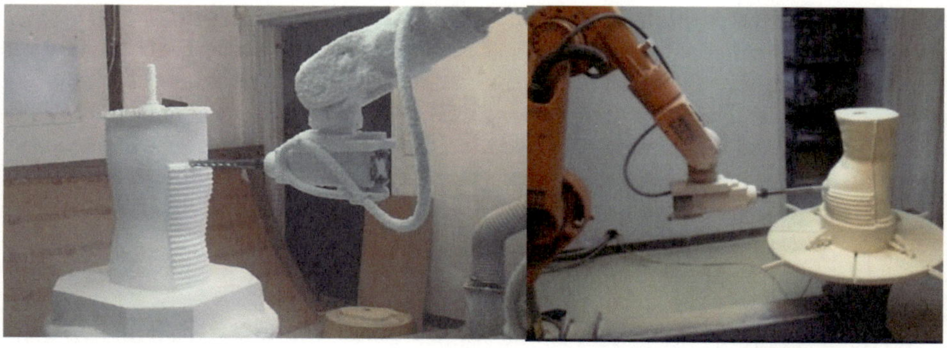

Figure 9.
Robot-assisted spinal orthosis molds production.

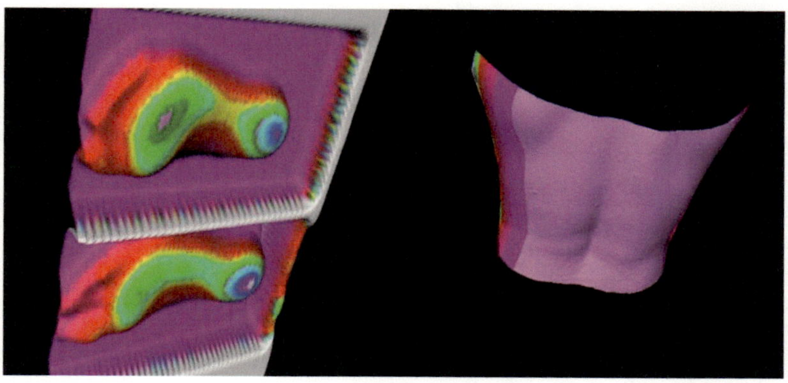

Figure 10.
Results of 3D scanning of foot foam box impressions and patient's back.

The resulting surfaces serve as the basis for design that will be modified or, upon superimposition onto the surface of the template, will be modified in terms of defining the corrective elements that give the functionality to orthosis. Upon completion of the design, the tool path is generated and transferred to the CNC machine tool [3, 12].

The use of CAD/CAM systems has not yet become widespread. There are reports in the literature that digital technology is currently in use in approximately a quarter of orthotic laboratories in developed countries, with the primary reasons being price, impractical performance, inconsistency, and the size of production systems. CAD/CAM technology is still a relatively new approach to O&P practice. The use of this technology has aroused great interest, but experiences are still being gathered and the approach is still evolving. Initial expectations from such systems were quite high in terms of increasing production, reducing manufacturing time, and obtaining high-quality products with a high degree of completeness, thus significantly reducing the required manual work. However, in a significant number of commercially available systems, this was not the case and such a system needs to be significantly refined to be adjusted for application in O&P. With implementation of new technologies, there is usually an initial doubt and hesitation in its application, especially if it is fundamentally different from the usual principles and if it requires the possession of additional knowledge in the IT and technical fields. However, it is undoubted that the cost of the initial investment in such production systems is the leading reason why they have not been widely implemented in practice. Other reasons include the problem of implementing this technology in production environments with limited space, impractical performance, and excessive complexity in use. From the above, it is concluded that any technological advancement of existing systems will be well received in practice [6, 13, 14].

Even a very detailed analysis of the available literature does not provide specific information on the important technical characteristics of the machines used in orthoses manufacture, through which it is possible to look at the design needs and modes of operation of the machine in the processing of different types of materials, and to optimize the production process. In practice, such data are generally obtained by trial and error [15].

It is also very rare to find data on the application of machines based on parallel kinematics [16].

4. Machine tools in orthotics

The classic three-axis CNC machine tools are characterized by rigidity and the possibility of heavy loads during operation.

The kinematic characteristics of most of the available machines, which are commonly used to make orthotic molds, assume mainly three- or four-axis milling CNC machines.

In general, the following characteristics are expected from machines used in the orthotic industry:

- High machining speed and high feed rate speed
- Adequate quality of the machined surface
- High repeatability
- High flexibility of use in terms of the possibility of making molds for orthoses and prosthetic allowances of different uses
- Appropriate work space for making the molds of maximum expected dimensions
- Adequate rigidity
- Energy efficiency
- Quiet operation
- Compactness and the least possible dimensions
- The possibility of extending the functionality with additional equipment for performing measurements

These aspects are further complemented by economic issues, which must certainly take into account the fact that the final performance of the machine must be satisfactory since it is a major requirement for its implementation in practice. It is also imperative that the machine is safe for operation and that the necessary safety and technical conditions are met.

5. Machine concept and realization

The most represented machines in orthotic practice are the three-axis CNC milling machines, which are generaly less demanding versions of standard metal cutting machines. When a specific machine for orthoses manufacture is being developed, this activity is complex and involves consideration of many aspects from the medical and technical side.

Methods for achieving technical realization of the machine and related software support include:

- Defining the required dimensions of the machine workspace, machine's external dimensions, its characteristics defined by the needs of the profession

- Development of a method for the solution of machine kinematics (solution of direct and inverse kinematics for the selected configuration), with the development of a software simulator to validate the kinematic model under virtual conditions

- Development of a software support that enables the generation of compensated tool paths

- Development of a simulation program with the possibility of realistic spatial simulation of machining the blank

- Development of the machine prototype, including all necessary stages: design, definition of drive and control components, simulation, preparation of project documentation, procurement of parts and all necessary materials, assembly of the machine and its start-up, implementation of necessary tests and corrections in order to ensure minimum technical requirements and meet all safety needs

After the machine realization follows the implementation of activities aimed at determining the correctness of the assumptions that the application of this solution contributed to the advancement of the orthoses manufacturing, accelerated and simplified production, reduced the need for human physical labor, and generally improved the current state of technology.

Except for the use of industrial robotic arms in mold machining, machines with five degrees of freedom are rarely encountered in orthotic practice.

In regard to experimental purpose, one also considers designs of machines that are based on parallel kinematics, which are expected to give some answers to problems arising through the application of conventional CNC machine designs. As the orthoses manufacture requires a high productivity to meet the requirements set by practice, the machining of materials in terms of generating complex surfaces, the production of large models (we often encounter the need to create models with a height of more than 800 mm and radii larger than 250 mm), there is a need for a technical solution that will improve and integrate these characteristics, as this is a disadvantage of classic designs.

This requirement is complemented by the fact that the materials used to make the orthoses or their molds (or base materials for the orthoses) are blanks made of cork, EVA, expanded polyurethane, polystyrene, MDF, or, less frequently, wood. All they belong to relatively soft materials, what significantly reduces the requirements on rigidity and power of needed for CNC machine tools applied in metal cutting.

Blanks typically come as templates in the form of blocks, plates, rollers, truncated cones, and shapes for making specific orthoses or their molds, which by shape and dimensions are the closest to the part of the body to which they are applied. Predefined shapes, templates, significantly increase process efficiency by reducing machining time and volume of the removed material.

The use of robotic arms for purposes of machining has its greatest value in great flexibility in terms of designing different types of molds and manipulation of five, or more, degrees of freedom and a large working volume.

The designs of the machines used in orthosis production are largely defined by the type of orthoses or molds for the orthoses being made.

In addition, the technical requirements are also conditioned by the geometric complexity of the orthosis surface, the materials that will be processed, and the time required to complete the product.

From all of the above, it is evident that the production of orthoses and related products has specificities that are different from the usual use of CNC machine tools in the industry.

Although these technologies are also used for the production of "off the shelf" orthoses, in most cases individual products or molds for their production, after being machined, will require significant additional human labor (despite the fact that the manufacturing process is automated) in order to be ready for application. The orthoses are individual products with strong request for short production time. Therefore, in the part of the production process that is automated, all the operations should be as simple as possible to implement. The data defined in the design phase (type of material, surface quality requirements, cutting tools) allow the possibility to predefine a set of machining parameters that could be selected in the design phase. The tool path generation should also be simplified since it must be repeated when creating each individual item, which is a time-consuming process. Consequently, applied software should have a high degree of automation and ability to generate optimized trajectories especially for three-axis CNC machine tools with one rotational axis, since they have a problem of hidden areas, which, due to limited kinematic performance of such machines, cannot be machined. Since orthotic blanks of large dimensions and various shapes are often used in the manufacture of orthoses molds, in tool trajectory generation special attention is given to rough machining, which typically removes a large amount of material. This raises the next important question, the elimination of dust and chips created during machining, and the big problem we face in practice is their quantity and properties.

Although orthotics use materials that have been tested and certified for use, and these are usually polymers such as polyurethane, polystyrene, PP, PE, EVA (ethyl vinyl acetate), UHMWPE, etc. due to the machining process itself, the separated particles (chips) are electrically charged and attach to the machine components, which create a particularly big problem if the guides and other moving components are not adequately protected.

A further related problem lies in the fact that materials used are often porous, and internally still contain a significant amount of volatile chemicals that are released by machining, increasing their concentration in the machine's workspace. In addition to adversely affecting the health of employees, these substances, by depositing on moving parts of the machine, can create a film that damages the sealing elements in the pneumatic components, or dissolve dust particles and create a solid deposit that compromises the operation of the bearings.

Therefore, it is necessary to provide systems that allow efficient evacuation and safely dispose of the particles generated by machining, since these are lightweight materials and particles of this dust, even at very low air flow, fly away from the surface and create high concentrations in space. Given these facts, it is highly desirable that these machines have a closed cabin and a good chips and dust collection system.

Application of cooling is less commonly used in such systems. However, with some materials and when requiring the use of specific tools to achieve high surface quality (often in the case of production, EVA insoles that do not have an extra layer of cover material, or when making functional orthoses made of polypropylene), increase of tool heating can lead to the insole material melting or even igniting.

Although such situations can be substantially reduced by selecting the appropriate values of the cutting parameters, compressed air is directed to the tool for cooling purposes. It cools the tool, but its application is also useful in blowing chips. This is especially important for contouring performed to define the external orthosis boundaries.

An important issue we often face in the use of CNC machine tools in the manufacture of orthoses is related to the clamping of the blank. As mentioned above, these are lightweight materials that need to be clamped as quickly as possible in the machine's workspace. In most cases, the use of clamping devices, such as those used in clamping metal blanks, is not considered, with the exception of the manufacture of rigid molds (e.g., MDF material for the purpose of making molds for deep-profile plastic foot orthoses, or making molds for head orthoses when there is a high requirement for the accuracy of the resulting surface). Most often we come across simple manual clamping devices that are easy to manipulate and adjust the clamping force. This however is not simply applicable to all types of materials, which is particularly the case with materials that are softer and more elastic, where it is not easy to exert an adequate clamping force without significant deformation. The double-sided adhesive tape is commonly used to temporarily affix the blank to the machine's work table. Although quality material placement is achieved in this way, the process takes a considerable amount of time before and after completion of the machining, and in most cases the application of double-sided adhesive tape is a significant financial expense (**Figure 11**).

For machines of larger dimensions, a vacuum table is commonly used for clamping, and its application greatly simplifies the process of setting and fixing materials in the machine's work space (**Figure 11**).

When using CNC lathes or multitasking machine tools we also encounter designs of clamping devices in which a shaft is fixed to the rotary axis, with a longitudinal axis collinearly positioned to the axis of rotation.

The blank is attached to this shaft. To clamp such a blank, a nut with a thrust plate is used, which sometimes contains shallow, sharp pins that stick into the material, further securing it. Typically, there are designs of the machines where this shaft is mounted horizontally or vertically. In vertical designs, those with special

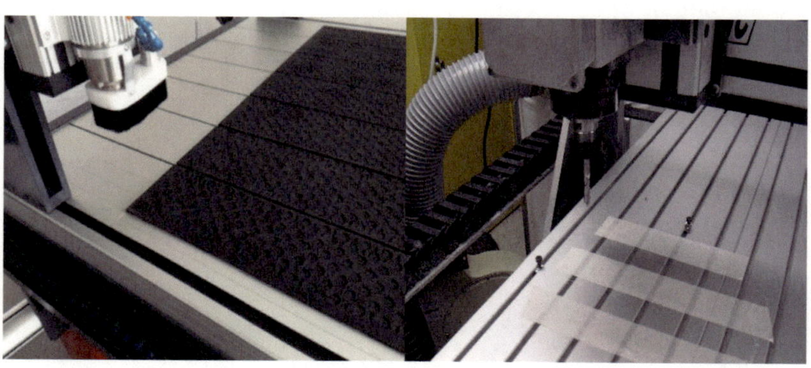

Figure 11.
Vacuum table and double-sided tapes for workpiece fixation to machine table.

assemblies for holding the blank are encountered, which additionally secure the blank from above and prevent machining vibrations.

Since the cutting forces in machining orthoses or their molds are incomparably less than those used in metalworking, the power requirement of the spindle motor often does not exceed 3 kW, and given the type of material being machined and the quality of the machined surface, there is a need for high rotational speed. It is desirable to achieve 20,000 RPM and above.

The most commonly used tools for machining of the material include the end mill, ball nose mill, and burr design mill. Typically, milling tools with high helix angles are used, which results in better particle removal and maintains a lower local temperature. The use of burr mills is found in cases where it is necessary to achieve high quality of the machined surface in the shortest possible time or when very soft materials are being machined. This entails the need for a very high spindle speed (≥30,000 RPM) (**Figure 12**).

When making orthoses, the most often used are the milling cutters with a diameter of 6–12 mm, while in the design of molds for body orthoses we encounter milling cutters with diameters up to 30 mm. Such milling tools can be longer than 300 mm, which places special requirement on the design of the tool clamping system.

In the design of such machines, it is expected that the access to the workspace and placement of blanks be extremely easy, since they can be larger in size and it is desirable to affix them to the work table in a way that excludes high clamping forces. In addition, often during machining there is a need to create a mold or product with complex geometry. In doing so, the possibility of fracture of the blank or its separation from the clamping device is increased (high centrifugal forces caused by unbalanced masses), which creates the need for rapid shutdown of the machine and easy access to the machine's working space.

The design of the machine is also expected to be as compact as possible; as such machines are most often installed in smaller sized orthotic laboratories that adapt to new tasks, causing machines and furniture to be moved frequently.

There is also a requirement for the machine to be very quiet, as such workplaces are often located near or within healthcare facilities where the general presence of a noise source is highly undesirable.

In addition to the manufacture of orthoses themselves, CNC machine tools with modifications are also used for other purposes.

In particular, by using them with a laser module or a tangential knife head, blocks of EVAe, cork, PP, HMWPE, UHMWPE, polyurethane, and other materials are cut into predefined shapes or contoured and allows other necessary cutting actions to be performed at high speeds.

In addition, in practice it is common to use mounted scanning module (camera with laser module or other vision devices) to scan surfaces (e.g., to scan foot

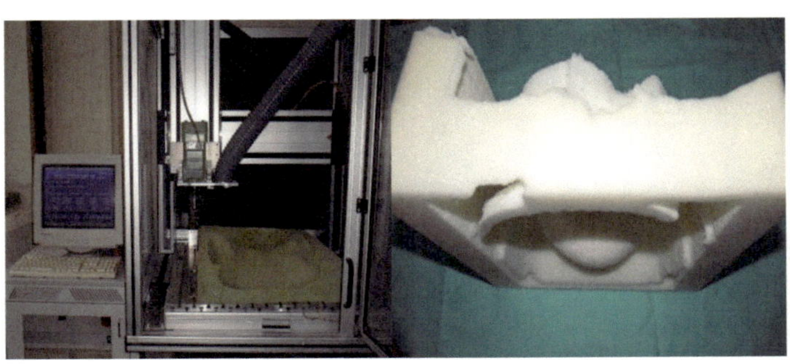

Figure 12.
Milled sponge back support and soft facial mask.

impressions in polyurethane foam), especially in machines with large work surfaces (e.g., three-axis CNC milling machines with large X and Y strokes).

Although rarely used, tool changer applications are very useful, especially in the manufacture of orthopedic insoles, in which ball nose cutters are used for machining the main surface, while end mill cutters with smaller radii are much more desirable for contouring.

Making each type of orthosis has its own peculiarities. The same product can be manufactured in a number of ways, although CNC machines can be used in all cases. Ideally, there would be a universal machining system, which is not the case at least for now.

In practice, the closest example of this might be the industrial robotic arm with at least five degrees of freedom, but for the time being there are some problems connected with use of this technology, which is why this manufacturing system is not overly present in orthotics.

6. Basic implementations of machine tools used in orthotics

In O&P practice, three-axis machine designs are used in most cases. Usually, these are standard industrial milling machines that are modified or upgraded by a simple operation for the needs of the specificity of such production (**Figure 13**).

The leading reason for the use of such implementations lies in the fact that generally in orthotics, CNC machine tools are most commonly used to make foot orthoses, primarily orthopedic insoles.

The use of three-axis CNC machine tools satisfies most of the requirements and by using them it is possible to make an orthosis or mold for it with a high degree of completeness, that is, with little need for additional modifications. Creating a pair of orthopedic insoles usually takes 6–16 min. The dimensions of these machines range from the very small ones on which it is possible to make a pair of inserts to those with a large work area commonly used in large laboratories. Three-axis CNC milling machines are also used for other purposes, such as machining of molds for craniofacial orthoses and masks.

Also, it is possible to make more complex molds, whereby the blank of known dimensions is after machining of one side rotated, for example, by 90° and fixed again on the work surface(**Figure 14**). If the dimensions of the working volume allow it, in this way it is possible to create molds for orthoses for various purposes. However, this procedure requires careful positioning and is time-consuming.

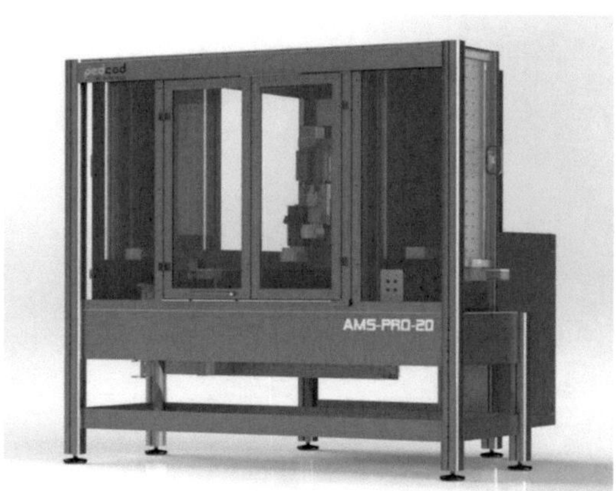

Figure 13.
Modern three-axis milling machine for foot orthoses production (courtesy of PedCAD GmbH).

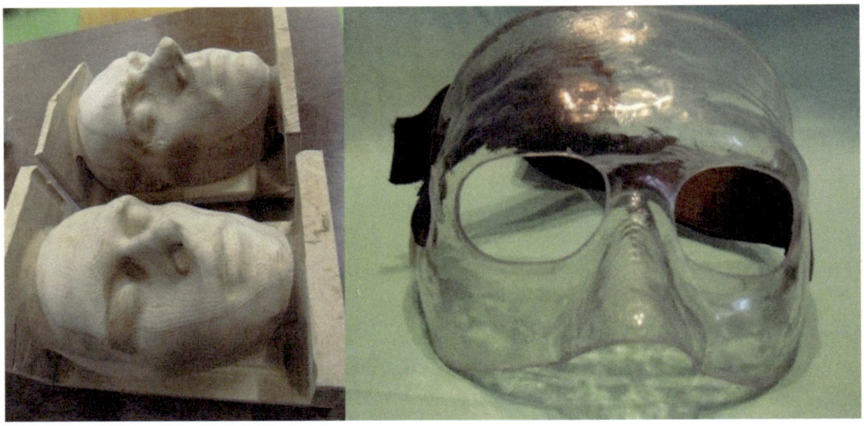

Figure 14.
Molds made on three-axis milling machine and custom protective facial mask.

Upgrading the three-axis CNC milling machine with the fourth rotary axis gives a machine more functionality [11].

In practice, we come across versions that are simply modified for specific needs. For example, for the purpose of machining orthopedic insoles, an accessory with a flat work surface is used on which the blanks are affixed, while for the manufacture of molds for spinal orthoses, a machine tool accessory is used in which, instead of a table, there is a fourth rotary axis.

Generally, the rotary axis is a commonly used component in such machines, with the exception of those for making orthopedic insoles. The reason is that the molds used to make the orthoses resemble the part of the body for which the orthoses are made and are created by machining a blank that rotates around a defined axis. This enables the formation of a continuous surface and complex shapes.

Such machines include CNC lathes or multitasking machine tools. Typical designs are those with a rotary axis set vertically or horizontally. They are usually used for machining blanks from expanded polystyrene or polyurethane. They are characterized by a large working volume (they are used to make molds for all types of orthoses, and molds for footwear; they are not particularly suitable for the production of orthopedic insoles, although they are also used in this regard), high efficiency, and an extensive particle removal system during machining. The average

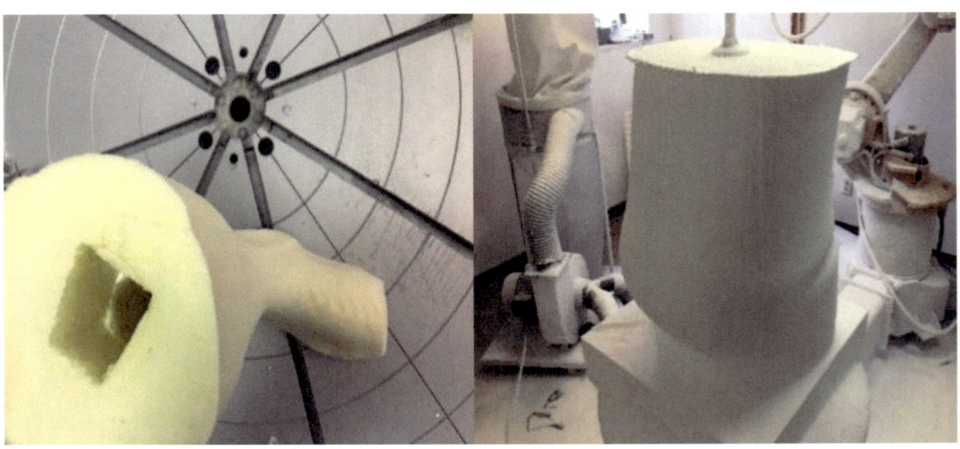

Figure 15.
Molds produced from expanded PU foam.

production of molds for orthoses of larger dimensions on standard versions of these machines takes between 25 and 60 min (**Figure 15**).

For these purposes, standard versions of machines with rotary axes are rarely used, since in most cases they are intended for metalworking; so, they have a relatively small working space and accompanying components are not intended for the dust produced during the processing of the aforementioned materials.

Application of CNC machine tools with five simultaneous axes in medical field is mostly present in manufacture of mold and dentures in dentistry [17, 18].

Such systems enable the production of all types of molds and orthoses, possess exceptional dexterity, and usually have a large working space. The problems associated with them concern their control, which is more complex than for the usual Cartesian machines; rigidity; speed; and several other issues. Some specific examples of their use in orthotic practice include performing specific automated actions such as applying adhesive by spraying, dyeing, cutting material using a laser module, and also forming the orthosis shape after adding a 3D printing head.

In use or in the form of realized prototypes, there are special machine designs that, by their construction and other characteristics, contribute to individual functionality, making such a machine more efficient, of more appropriate dimensions, or possessing other features that make it more applicable in O&P practice.

Among the many, mentioned here is a special design of an industrial prototype of an orthopedic insole making machine (IPASIOU Project, Faculty of Mechanical Engineering and Naval Architecture Zagreb, Croatia) [19], characterized by a high degree of automation, which, with the automated digitization of both feet, enables an automated process of designing orthopedic insoles, including the top and bottom sides of the insoles (enabling footwear application without the need for additional modifications), and an automated tool path generation for a seven-axis machine with four Z axes (two pairs of collinearly coupled Z axes, which are autonomous and enable simultaneous processing of four different surfaces). As a result of carrying out this procedure, a pair of individually made insoles are prepared that are ready for application in footwear without further modifications, and the whole process of making a pair of such insoles from digitizing the feet to extracting the workpiece from the workspace takes 8–12 min (**Figure 16**).

Another interesting example of a machine with special functionality, suitable for O&P, is found in a specially made CNC machine (Ortoflex, Faculty of Mechanical Engineering SlavonskiBrod, Croatia) [20].

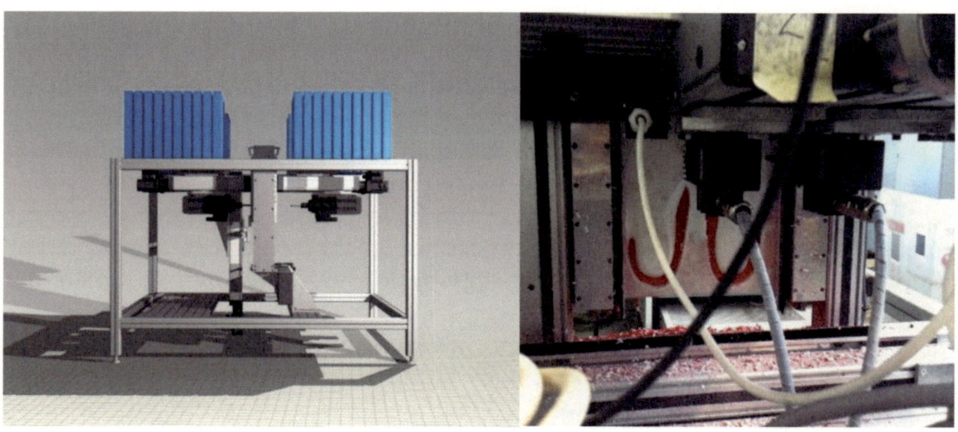

Figure 16.
IPASIOU system.

Figure 17.
Ortoflex five-axis milling machine based on parallel kinematics.

This machine is assembly based on a parallel kinematic design where the platform moves in the X, Y, and Z directions (**Figure 17**). The machine also includes two rotary axes. The primary purpose of the machine is to make molds for spinal orthoses, but it can also be used to make other molds as well. Its special feature is very high speed of movable platform on which the main spindle is located, and the possibility of machining shapes of complex geometry that cannot be easily realized on machines with four axes or fewer. Both projects, IPASIOU and Ortoflex, are the result of research aimed at improving the state of the art in the production of orthoses, in which the authors of this text participated. Of course, there are numerous other specific machine designs.

7. Conclusion

This review discusses the use of CNC machine tools in the production of orthoses and molds for their manufacture. In this area, the application of such systems becomes a standard.

A similar approach is used in the design of individual prosthetic devices, although there are significant differences in the applied materials and functionality of such aids and the methodology of their manufacture and application, which is a much broader context, so their manufacturing is not elaborated in this review.

There is also a range of other related products, the production of which is very similar to the production of orthoses such as individually made wheelchairs, inserts for standard seats, individually made mattresses, pillows, etc. where the use of CNC machine tools as well as industrial robots has enabled quick and precise workmanship.

Although there are still many open questions why the use of CNC machines in the O&P industry is not high, it is expected to become a standard manufacturing technology soon. The use of digital manufacturing technologies has radically changed the orthotic practice; made it simpler, better, faster, more accurate; and has introduced many other benefits. In contrast to the classical approach, it is also possible to exchange information and experience quickly, which contributes to the improvement of the methodology of orthotic design, to further optimization of the production process and the resulting savings. Certainly, the biggest beneficiary of this approach is the patient who gets more advanced orthosis in a shorter time.

Although competing technologies such as additive manufacturing are emerging, the advantages of the application of CNC machine tools in the manufacturing of orthotics will not be easy to achieve. Moreover, with the advancement of technology, it is expected that machine tools for this purpose will become more widely available and find wider application in the O&P field.

Author details

Karlo Obrovac[1*], Miho Klaić[2], Tomislav Staroveški[2], Toma Udiljak[2] and Jadranka Vuković Obrovac[3]

1 Ortogen d.o.o., Zagreb, Croatia

2 Faculty of Mechanical Engineering and Naval Architecture, University of Zagreb, Zagreb, Croatia

3 Pentatlon d.o.o., Zagreb, Croatia

*Address all correspondence to: karlo.obrovac@zg.htnet.hr

References

[1] Obrovac K, Raos P, Galeta T, Nižetić J, Mutka A, Vuković Obrovac J, et al. A new approach to the design of a CNC machine for making orthotic molds. Tehničkivjesnik. 2018;**25**(Supplement 2):460-465. DOI: 10.17559/TV-2018051119

[2] Wohlers TT, Gornet T. History of Additive Fabrication (Part 2). Time-Compression Technologies; 2008. https://wohlersassociates.com/articles2.html

[3] Kennedy S. The Future of O&P: A Practitioner's Perspective. The O&P EDGE. 2009. Available from: http:\\www.oandp.com/edge

[4] O&P in the Digital Age—Digitalisation is the Highlight Subject at OTWorld 2018, OTWorld. In: International Trade Show and World Congress, 15-18 May 2018. Available from: https://www.ot-world.com/press-releases/press-event-otworld-2018-press-folder/797673

[5] Chen D. Orthotics and Prosthetics Go Digital. Medical Design Technology. 2009. Available from: https://www.mdtmag.com/article/2009/05/orthotics-and-prosthetics-go-digital

[6] Sanders JE, Severance MR, Myers TR, Marcia CA. Central fabrication: carved positive assessment. Prosthetics and Orthotics International. 2011;**35**(1): 81-89. DOI: 10.1177/0309364610394476

[7] www.pedcad.de [Accessed: 15 March 2020]

[8] https://www.provel.us/ [Accessed: 15 March 2020]

[9] http://rodin4d.com/en/Products/manufacturing [Accessed: 15 March 2020]

[10] http://roboticom.it/brand/ortis/ [Accessed: 15 March 2020]

[11] https://vorum.com/op-carvers/ [Accessed: 15 March 2020]

[12] Bílek O, Jakub J, David S. 3D data acquisition and CAD/CAM systems for CNC manufacturing of artificial limbs. In: Latest Trends on Systems—Volume I; Proceedings of the 18th International Conference on Systems (part of CSCC '14); Santorini Island, Greece; 2014

[13] Lilja M, Oberg T. Volumetric determinations with CAD/CAM in prosthetics and orthotics: Errors of measurement. Journal of Rehabilitation Research and Development. 1995;**32**(2): 141-148

[14] Smith DG, Burgess EM. The use of CAD/CAM technology in prosthetics and orthotics—Current clinical models and a view to the future. Journal of Rehabilitation Research and Development. 2001;**38**(3):327-334

[15] Agarwal N. Surface roughness modeling with machining parameters (speed, feed & depth of cut) in CNC milling. MIT International Journal of Mechanical Engineering. 2012;**2**(1):55-61

[16] Merlet J-P. Parallel Robots. 2nd ed. Sophia-Antipolis, France: Springer/INRIA; 2006

[17] Rudolph H, Luthardt RG, Walter MH. Computer-aided analysis of the influence of digitizing and surfacing on the accuracy in dental CAD/CAM technology. Comput Biol Med. 2007;**37**:579-587. DOI: 10.1016/j.compbiomed.2006.05.006

[18] Lebon N, Tapie L, Duret F, Attal JP. Understanding dental CAD/CAM for restorations—Dental milling machines from a mechanical engineering viewpoint. Part A: Chairside milling machines. International Journal of Computerized Dentistry. 2016;**19**:45-62

[19] https://patents.google.com/patent/WO2014049379A1 [Accessed: 15 March 2020]

[20] http://ortoflex.sfsb.hr [Accessed: 15 March 2020]